SRA Snapshots
Video Science

Mc Graw Hill **SRA**

Columbus, OH

About the Front Cover

Ocelots (*Felis pardalis*) live throughout Latin America and are found as far north as Texas. Their diverse habitats include rain forests, mountain forests, thick bush, semi-deserts, coastal marshes, and even river banks, but they are never found in open country. Ocelots can climb trees but they usually remain on the ground. Ocelots are known for their striking spotted coats, which provide camouflage for the cats as they hunt. They are skillful hunters that feed mostly on small rodents. Ocelots hunt at night and have no problem seeing their prey because their eyesight is much better than humans' eyesight.

Female ocelots carry their young longer than most cats, and they do not have very many cubs at a time. Although they can live as long as ten years in the wild, ocelot populations are not rapidly increasing. Ocelots are listed as threatened or endangered throughout their habitat.

SRAonline.com

 SRA

Send all inquiries to:
SRA/McGraw-Hill
4400 Easton Commons
Columbus, OH 43219-6188

0-07-609679-3
978-0-07-609679-4

2 3 4 5 6 7 8 9 BCM 13 12 11 10 09 08

The McGraw·Hill Companies

Contributing Authors

Kate Boehm Jerome
Science Writer/
Curriculum Specialist
Charleston, SC

Julie Osteig
Elementary Science Coordinator
Charlotte Mecklenburg Schools, NC

Debra Young
Retired Technology Teacher, Consultant
School District of Palm Beach County, FL

Academic Consultants

Vocabulary Development

Michael F. Graves, PhD
Center for Reading Research
University of Minnesota
St. Paul, MN

English Language Learner Support

Suzanne Panferov, PhD
Director of the Center for English as a Second Language
University of Arizona
Tucson, AZ

Content Area Reading Development

Doug Fisher, PhD
Professor of Language and Literacy Education
San Diego State University
San Diego, CA

Science Content Development

William C. Kyle Jr. PhD
E. Desmond Lee Family Professor of Science Education
University of Missouri–St. Louis
St. Louis, MO

Marc Branham, PhD
Department of Entomology and Nematology
University of Florida
Gainesville, FL

Testing Support

John Zbornik, PhD
Psychologist
Lakewood City Schools
Lakewood, OH

Program Reviewers

Kay Daughtry
Science Teacher
Elizabeth Traditional Elementary
Charlotte, NC

Larry Hohman
Science Teacher
Grizzel Middle School
Dublin, OH

Stephen Houser
TD Coordinator/Science
Providence Spring Elementary
Charlotte, NC

Rebecca Johnson
Science Consultant
Sioux Falls, SD

Katherine Kendall
Science Educator
Ashland City, TN

Anita Shively
Elementary Teacher
Smith Elementary School
Delaware, OH

Sharmagne Solis
Science Coordinator
Galaxy Elementary School
Greenacres, FL

Contents

CHAPTER 1 Living Things

CHAPTER 2 Ecosystems

CHAPTER 3 Humans

Contents

Features

SCIENCE KNOW ZONE

Standardized Test Practice

Science Handbook

How do I study science?

Hello, Webster here, your guide to all things science. As you go through **SRA Snapshots Video Science**™, I will be here to help you. This program has two main parts: the videos and the book. Together, they help you learn important science ideas and vocabulary.

To succeed in **SRA Snapshots Video Science**™, you only have to follow these three easy steps: **See it! Read it!** and **Learn it!**

See It!

Your first job is to carefully watch the videos. They will teach you all you need to learn. And they're fun to watch. I should know, because I'm in them!

Read It!

After watching the video, review what you've learned by reading this book. It reviews all the important ideas and vocabulary words that you need to know. I'll be along to help.

Learn It!

This program lets you show off what you have learned. Each video and book lesson ends with activities that allow you to Show What You Know!

Check Your Understanding

Show What You Know

Main Ideas: Write the answer to each question.
1. Why do some organisms avoid competition?
2. What are some interactions that help organisms survive?
3. How do migration and barriers affect some organisms' chances for survival?

Competition

See Science by watching the videos. Each video lesson has all the important points you need to know. I'll drop in to help you with the definitions of your vocabulary words.

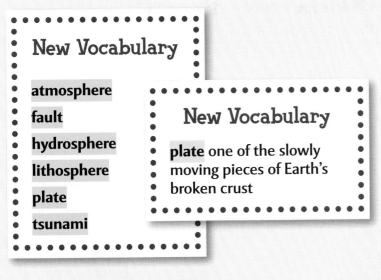

New Vocabulary

- atmosphere
- fault
- hydrosphere
- lithosphere
- plate
- tsunami

New Vocabulary

plate one of the slowly moving pieces of Earth's broken crust

Every time you watch a video lesson, you get a closer look at all the new vocabulary terms. You can even hear the vocabulary terms in English or Spanish in the video glossary.

The videos introduce all the main ideas you need to learn. Each main idea is covered in its own video segment.

At the end of each video segment, the main idea is reviewed for you.

Soon You'll Know

Main Ideas

1. How Earth and its atmosphere are divided into layers
2. How Earth's crust moves
3. What kinds of changes happen when plates move

Now You Know

Main Idea 1: Earth's Layers

Scientists have divided Earth and its atmosphere into many layers including the atmosphere, hydrosphere, and lithosphere.

Now You Know

Main Idea 2: Plate Movement

Earth's crust is separated into plates that are slowly and constantly moving.

Now You Know

Main Idea 3: Volcanoes and Earthquakes

Plate movements cause sudden changes such as earthquakes, and gradual changes, such as mountains and valleys.

What's happening in Earth's crust?

I'll give you a question at the start of each lesson. You should be able to answer it by the end of the video.

How do I study science?

Read science by using your book. You've already had your first experience with it. You're reading it now! You'll notice that the book and the videos have a lot in common. They were made to go together.

Remember this page from the video? Here it is again in your book. Each lesson has a page like this. The title, question, vocabulary, and main ideas for the lesson will appear on this page.

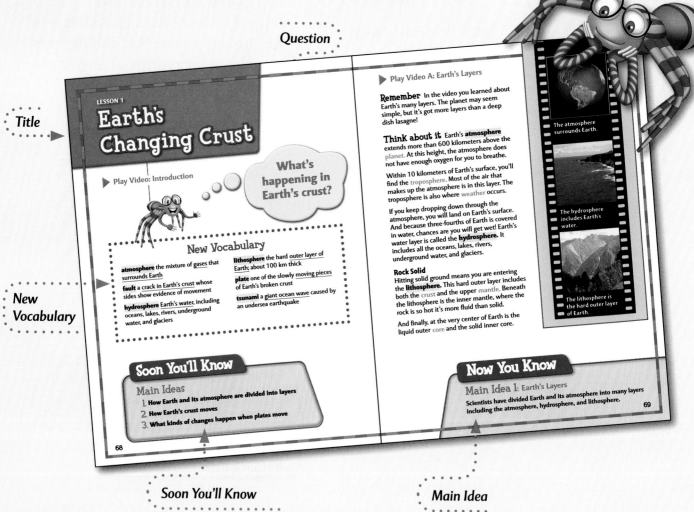

Question

Title

New Vocabulary

Soon You'll Know

Main Idea

(content shown in the book image)

▶ Play Video A: Earth's Layers

Remember In the video you learned about Earth's many layers. The planet may seem simple, but it's got more layers than a deep dish lasagne!

Think about it Earth's **atmosphere** extends more than 600 kilometers above the planet. At this height, the atmosphere does not have enough oxygen for you to breathe.

Within 10 kilometers of Earth's surface, you'll find the troposphere. Most of the air that makes up the atmosphere is in this layer. The troposphere is also where weather occurs.

If you keep dropping down through the atmosphere, you will land on Earth's surface. And because three-fourths of Earth is covered in water, chances are you will get wet! Earth's water layer is called the **hydrosphere**. It includes all the oceans, lakes, rivers, underground water, and glaciers.

Rock Solid
Hitting solid ground means you are entering the **lithosphere**. This hard outer layer includes both the crust and the upper mantle. Beneath the lithosphere is the inner mantle, where the rock is so hot it's more fluid than solid.

And finally, at the very center of Earth is the liquid outer core and the solid inner core.

The atmosphere surrounds Earth.

The hydrosphere includes Earth's water.

The lithosphere is the hard outer layer of Earth.

LESSON 1
Earth's Changing Crust

▶ Play Video: Introduction

What's happening in Earth's crust?

New Vocabulary

atmosphere the mixture of gases that surrounds Earth

fault a crack in Earth's crust whose sides show evidence of movement

hydrosphere Earth's water, including oceans, lakes, rivers, underground water, and glaciers

lithosphere the hard outer layer of Earth; about 100 km thick

plate one of the slowly moving pieces of Earth's broken crust

tsunami a giant ocean wave caused by an undersea earthquake

Soon You'll Know

Main Ideas
1. How Earth and its atmosphere are divided into layers
2. How Earth's crust moves
3. What kinds of changes happen when plates move

68

Now You Know

Main Idea 1: Earth's Layers
Scientists have divided Earth and its atmosphere into many layers including the atmosphere, hydrosphere, and lithosphere.

69

At the end of each page, the main idea for the lesson is explained. You'll also notice that some words are **highlighted** or appear in blue. Highlighted words are **new vocabulary**. The blue words are vocabulary words from other chapters. You can check out the Glossary to find out what these words mean.

Learn It!

Learn science by testing yourself. After watching the video and reading the book, you should know lots of new things. Test yourself on what you learned by using the MindJogger game in the video and the review pages at the end of each lesson. Improve your test-taking abilities by using the tips in the Standardized Test Practice pages.

The lesson review pages give you a chance to review vocabulary again.

Quiz yourself to see how well you remembered the main ideas from the lesson.

Build Your Vocabulary

Vocabulary Review
Use the word bank to complete each statement.

1. The layer of gases that surrounds Earth is called the _____.

2. All of Earth's water forms the _____.

3. A crack in Earth's crust whose sides show evidence of movement is a _____.

4. One of the slowly moving pieces of Earth's broken crust is called a _____.

5. The hard, outer layer of Earth's surface is called the _____.

6. A _____ is a giant ocean wave caused by an undersea earthquake.

> atmosphere
> fault
> hydrosphere
> lithosphere
> plate
> tsunami

Word Study: Word Roots
Many parts of science words come from other languages. Those parts are called roots.

The root *sphere* means "ball" in Greek. The vocabulary word *atmosphere* uses this root.

| atmo | + | sphere | = | *atmosphere* |

1. Which other vocabulary words use this root?
2. Think of two more words that use the root word *sphere*.
3. Write the words and what they mean.

72

Check Your Understanding

Show What You Know

Main Ideas: Write the answer to each question.

1. What are the main layers that make up Earth and the area around it?

2. Why is Earth's crust slowly but constantly moving?

3. What kinds of changes can plate movement cause?

Critical Thinking

1. **Analyze** Explain why it's correct to say that forces under Earth shape its surface.

2. **Evaluate** Why do you think it's important to monitor earthquakes that happen under the oceans?

Math in Science

Calculate In an average day, about 8,000 earthquakes occur around the world. But less than one in 100, or $\frac{1}{100}$ of them, are strong enough to do real damage.

♦ Figure out what percentage of earthquakes cause damage.
♦ How many earthquakes would cause damage in an average year?

Process Skill Quick Activity

Infer The plate boundaries that circle much of the Pacific Ocean are known as the Ring of Fire. In fact, 75% of Earth's active and dormant volcanoes are located along the Ring of Fire.

Find a map of the Ring of Fire. List the continents that are affected by it. Tell why people who live in the area should be prepared, but not panicked.

73

Challenge yourself even further with the other activities on the page!

Standardized Test Practice 1

TIPS
Skipping Difficult Questions Answer as many questions as you can in the time you have. Skip difficult questions and come back to them later, if you have time.

Multiple Choice Practice Read each question. Choose the best answer.

1 Which of the following things is an organism?

A

2 Which environment might be home to a polar bear?

A one with lots of sand
B one with lots of snow
C one deep in the ocean
D one high in a tree

Each chapter gives you a new tip on how to be a better test-taker. Learn these skills and you'll soon be a test-taking pro!

Don't forget to check out the MindJogger Review in the videos!

Scavenger Hunt

Your **SRA Snapshots Video Science** textbook is full of great information. The trick is learning how to find that information when you need it.

This scavenger hunt activity will help you learn how your book is set up and figure out how to best use the text to help you.

1 How many chapters are in this book?

2 What do the colors behind the lesson titles mean?

3 What is the title of Chapter 4, Lesson 2?

4 If you want to quickly find the definition of the word *ecosystem*, where should you look?

5 How many vocabulary words are in Chapter 8, Lesson 3?

6 What is the Standardized Test Practice tip in Chapter 2?

7 Where can you find the answer to the Soon You'll Know prompts in each lesson?

8 If you want to quickly find the section on simple machines, where should you look?

9 What is the topic of the Science Know Zone in Chapter 7?

10 Why do some words in each lesson appear highlighted or in blue?

Living Things

- **Discussion** Life is leaping all around you! There are more living things in the world than you can count. What living things can you see here? Which things are not alive?

- **Critical Thinking** How do you tell the difference between a living and a nonliving thing?

1

Living Things Have Needs

▶ **Play Video: Introduction**

What do living things need?

New Vocabulary

environment the things that make up an area, such as land, water, and air

leaf a plant part that grows from the stem and helps a plant get air and make food

organism any living thing

oxygen a gas that is in air and water

root a plant part that takes in water and minerals

stem a plant part that supports the plant

Soon You'll Know

Main Ideas

1. What an organism is
2. What things plants must have to survive
3. What animals need to survive in their environment

Remember The video you just watched was about living things. You already know a lot about this topic. Why? Because you are a living thing too!

Think about it What do you do if you're hungry? You eat! What do you do if you're tired? You sleep! The fact is, you're pretty good at meeting your own needs.

You're not the only one who does this. All living things, which are called **organisms,** have needs. In order to live, organisms must find ways to meet their needs.

Living things are called organisms.

Organisms get what they need from their environments.

The **environment** that an organism lives in helps it meet its needs. Fish need to swim in water. Butterflies need to fly in air. Rich soil and good sunlight help flowers grow. So if their needs are met, organisms can survive.

Now You Know

Main Idea 1: Organisms

Organisms are living things that have needs.

Remember In the video you just watched you learned how plants meet their needs. They don't even have to move around to do it!

Think about it It's hard to believe that the tangled, dirt-covered **roots** of a plant are important. But roots do several things. They keep a plant in place. They also take in water and minerals that the plant needs to grow. Most roots grow under the ground.

Stems support a plant, and leaves make food. What do you think flowers do?

Plants need water, sunlight, and air.

Stems are important too. They support a plant and help move water and minerals around the plant. And of course, we can't forget about **leaves.** If a plant gets enough air, water, and sunlight, it can make its own food in its leaves. All together, roots, stems, and leaves help plants meet all their needs.

Now You Know

Main Idea 2: Plant Needs

Plants need sunlight, water, air, and minerals to survive.

Remember In the video you just watched you learned about animal needs. By the way, aren't you glad you don't have to eat as much as a rhinoceros does every day?

Think about it A crocodile dines on fish. A caterpillar chomps on leaves. Different animals eat different things. While they're eating, they all do something else too. They breathe! That's because animals need **oxygen,** which is a gas that is in air and water.

The oxygen animals need can be found in the air.

Animals need shelter.

Most animals can't live long without water.

Fish get oxygen from the water. You and many other animals get oxygen from the air. But, there's more to life than food and air.

Animals can't live without water. They also need shelter, which is a safe place to live and sleep. Food, oxygen, water, and shelter are important animal needs.

Now You Know

Main Idea 3: Animal Needs

Animals need food, water, oxygen, and shelter.

Build Your Vocabulary

Vocabulary Review

Use the word bank to complete each statement.

1. The part that helps a plant get air and make food is the _____.

2 Any living thing is called an _____.

3. The part of a plant that usually grows underground is a _____.

4. _____ is a gas that is found in both air and water.

5. A plant part that supports a plant is the _____.

6. Plants and animals get what they need in the _____ in which they live.

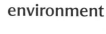

environment

leaf

organism

oxygen

root

stem

Word Play: Memory Sentence

Memory sentences can help you remember words or facts.

The first letter of each word in the sentence below can help you remember what plants need to live.

Many	**A**rtists	**W**ant	**S**cissors
Minerals	Air	Water	Sunlight

Create your own memory sentence using vocabulary words or ideas from the lesson.

Check Your Understanding

Main Ideas: Write the answer to each question.

1. What are organisms?

2. What are the four things that plants need to live?

3. What are the four things that animals need to live?

Critical Thinking

1. **Analyze** Why is a tree an organism but a rock is not?

2. **Synthesize** Describe what you would need to live on another planet.

Writing in Science

Write a paragraph Describe what a plant needs to stay alive.

- Use at least three science vocabulary words.
- Put details in your paragraph.
- Remember, a good paragraph includes:
  - a topic sentence
  - a body
  - a closing sentence

Process Skill — Quick Activity

Observe Look at the environment around you. Can you name the living and nonliving parts? Don't forget to think about the very tiny things in the environment.

Make two lists. Label one list "Living." Label the other list "Nonliving." List all the things you can think of that belong in those two lists. See which list is the longest when you are done.

Grouping Living Things

▶ **Play Video: Introduction**

How are living things grouped?

Soon You'll Know

Main Ideas

1. How scientists group living things
2. Which two groups scientists use to classify animals
3. How scientists group different plants

Remember In the video you just watched you learned that organisms are grouped. This is a good thing. If we couldn't group animals together, there wouldn't be any way to make sense of the millions of animals in the world!

Think about it The baboon mother feeds her babies milk. But the mother bird can't do this. That's because baboons and birds have different traits, or characteristics. Scientists use common traits to group animals together.

Animals are placed in groups according to like traits. These birds all have feathers.

All mammals have hair or fur.

A rattlesnake and a lizard might seem very different. But they both have scaly skin, so they belong to the reptile group. Amphibians, however, share different traits. They can live on both land and water. What about **mammals?** Well, if an animal has hair or fur and feeds milk to its young, it's a mammal!

Now You Know

Main Idea 1: Grouping Animals

Animals are grouped by their common traits.

▶ **Play Video B: Vertebrates**

Remember You just learned a basic fact in the video. In the animal world, you either have a backbone or you don't!

Think about it Reach around and feel those lumpy bones running down the middle of your back. Can you feel them? Those bones make up your backbone. Having that **backbone** means you belong to a special group. You're a **vertebrate!**

There are lots of different kinds of vertebrates in the world. Whales, bats, snakes, parrots, and goldfish are all vertebrates.

Animals with backbones are called vertebrates.

Animals without backbones are invertebrates.

Animals without backbones are called invertebrates. They make up an even larger group. One of the reasons this group is so big is because it includes all the crawling, jumping, flying **insects** of the world.

Now You Know

Main Idea 2: Vertebrates

Two big animal groups are those that have backbones and those that do not.

Remember In the video you learned that there are two main groups of plants. Isn't it amazing that it's the tiny seeds that make all the difference?

Think about it Green, leafy ferns don't have seeds. But green, leafy tomato plants do! In fact, a lot of the plants that you eat are **seed plants.** That means those plants grow from and produce seeds. But, it doesn't stop there. You can break plants down into even smaller groups.

Some plants don't make seeds.

Plants that do make seeds can be flowering or nonflowering.

Seed plants can be broken down into two groups. They are the nonflowering plants and the **flowering plants.** You might like this last group the best. That's because the seeds of flowering plants are often found inside the fruit you eat. Apples, oranges, and blueberries all come from flowering seed plants.

Now You Know

Main Idea 3: Plant Groups

Plants are grouped into two main groups: plants that make seeds and plants that don't make seeds.

Build Your Vocabulary

Vocabulary Review

Use the word bank to complete each statement.

1. A plant that produces seeds inside of flowers is a _____.

2. An animal with three body sections, three pairs of legs, and, usually, two pairs of wings is called an _____.

3. A _____ is a vertebrate with hair or fur that feeds its young with milk.

4. A long line of bones that runs down the back of some animals is a _____.

5. An animal that has a backbone is a _____.

6. Any plant that produces and grows from seeds is a _____.

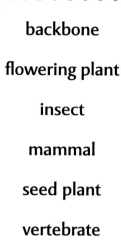

backbone

flowering plant

insect

mammal

seed plant

vertebrate

Word Study: Nonexamples

Examples show what something is. Nonexamples show what something is *not*.

Write an example and a nonexample for each of these words. The first one is done for you.

	Example	Nonexample
1. food	a. hot dog	b. rock
2. insect	a.	b.
3. mammal	a.	b.
4. flowering plant	a.	b.

12

Check Your Understanding

Show What You Know

Main Ideas: Write the answer to each question.

1. What do scientists use to put animals in different groups?

2. What is the one trait that separates the two main animal groups from each other?

3. What are the two main groups of plants?

Critical Thinking

1. **Evaluate** Why is it important for scientists to group living things?

2. **Apply** How are things grouped in your classroom?

Math in Science

Create a bar graph Survey your class to see what types of pets they have. Create a bar graph of the results.

♦ Make a tally sheet to record what kinds of pets students have.

♦ Group the pets into 5 categories: mammals, reptiles, birds, fish, and other animals.

♦ Graph the results in a bar graph.

♦ Use the graph to figure out the most popular pets in your class.

Process Skill Quick Activity

Classify Look at the pictures on the walls of your classroom. Can you think of different ways to classify these pictures? Think of things like shape, color, and size.

Make a T-chart that shows one way you could separate the pictures into two groups. For example, on one side you could list large pictures, and on the other side you could list small pictures.

Insects Rule!*

Has a mosquito ever bugged you? Has an ant ever come to your family picnic? Insects may not be big in size, but they are certainly huge in numbers. There are around 100,000 different kinds of insects in North America alone!

DID YOU KNOW? A honeybee's wings beat over 11,000 times per minute.

How Sweet It Is!

Honeybees live together in hives. Bees make honey for food. The honey is placed in little cells inside the hive. The cells are sealed with beeswax when they are full.

That's a Big One!

The Hercules beetle is famous for its size. This rain forest insect can grow to be 17 cm in length. It may give a scare, but this insect is not dangerous to people.

Critical Thinking

● How is an insect helped by its small size?

● Why do you think insects lay so many eggs?

Find Out More!

Research on the Web

Most insects have four stages to their life cycles. Find out about those stages.

The Insect Club

Which critter below is NOT an insect?

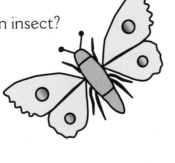

15

Life Cycles

▶ **Play Video: Introduction**

What happens during a life cycle?

New Vocabulary

fruit the part of a plant that grows around seeds

life cycle all the stages in an organism's life

metamorphosis a change in the body form of an organism

pollen a powdery material needed by the eggs of flowers to make seeds

reproduction the way organisms make more of their own kind

seedling a young plant

Soon You'll Know

Main Ideas

1. How living things grow and change
2. Some different animal life cycles
3. How a flowering plant starts its life cycle

▶ **Play Video A: Growth and Change**

Remember In the video you just watched you saw some cute little babies. How do you think those babies will change in the next five years?

Think about it It wasn't that long ago that you were a tiny baby yourself. But you know you're not a baby anymore! What happened? Like all organisms, you go through different stages of growth and change. This is called a **life cycle.**

Organisms have life cycles.   **Organisms grow and change during their life cycles.**

Think about how much a tree changes in its life cycle. It starts out as a tiny seed and grows into a tall tree. Now that's a big change! Animals grow and change too. A big lion starts its life cycle as a little cub, small enough for you to hold in your hands. As time goes by, plants, dogs, cats, and even tiny living things that you can't see grow and change during their life cycles.

Now You Know

Main Idea 1: Growth and Change

Living things grow and change in a life cycle.

Remember In the video you saw how a frog's life cycle is different from yours. That's a good thing. It would be pretty strange for you to live in the water as a baby and on land as an adult!

Think about it It's easy to imagine a baby horse growing into a tall, strong adult horse. But it's hard to picture a fuzzy caterpillar turning into a beautiful butterfly with wings. They don't look anything alike.

All animals have a life cycle.

Some animals go through metamorphosis.

A caterpillar goes through **metamorphosis** to become a butterfly. This means that big changes happen to its body form during its life cycle. Frogs go through metamorphosis too. A frog egg hatches into a tadpole. The tadpole soon grows legs. When the tail disappears, the frog's metamorphosis is complete. An adult frog is finally formed.

Now You Know

Main Idea 2: Metamorphosis

Some animals go through metamorphosis and some do not.

Remember In the video you saw many flowers. Their colors and sizes are different, but they all started out from the same thing–a seed!

Think about it Flowers are so pretty! But the way they look is not the only thing that makes them special. Flowers are important in the life cycle of a seed plant. Seed plants use flowers for **reproduction,** which is making more of that kind of seed plant.

Seed plants use seeds for reproduction.

Flowering seed plants need flowers to make seeds.

When a powdery material called **pollen** meets the eggs in a flower, new seeds start to form. A **fruit** protects the seeds while they grow. When seeds finally land in the right place to meet their needs, they germinate, or start to grow. New young plants, called **seedlings,** will soon appear. These will grow into adult plants that can make their own seeds and start the cycle again.

Now You Know

Main Idea 3: Flowering Plants

Flowering plants start their life cycles as seeds.

Build Your Vocabulary

Vocabulary Review

Use the word bank to complete each statement.

1. All the stages in an organism's life are called its _____.

2. _____ is the powdery material needed by the eggs of flowers to make seeds.

3. A change in the body form of some animals is called a _____.

4. A young plant is called a _____.

5. _____ is the way organisms make more of their own kind.

6. A plant grows its _____ around its seeds.

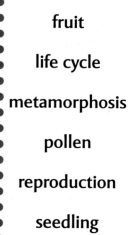

fruit

life cycle

metamorphosis

pollen

reproduction

seedling

Word Study: Word Roots

Many parts of science words come from other languages. These parts are called roots.

The root *morph* means "shape" in Latin. The vocabulary word *metamorphosis* uses this root.

meta + **morph** + **osis** = *metamorphosis*

1. Look in your dictionary for three more words that use the root *morph*.

2. Write the words and their definitions.

3. Also write a sentence using each word.

Check Your Understanding

Main Ideas: Write the answer to each question.

1. What do all living things go through as they grow and change?

2. Why do some animals look different at some stages in their life cycles?

3. How do flowering plants start their life cycles?

Critical Thinking

1. **Evaluate** Explain why it might be helpful to a plant for birds to eat its fruit.

2. **Analyze** What makes a frog's life cycle different from a cat's life cycle?

 Words in Science

Process Skill — Quick Activity

Academic Vocabulary Use the dictionary to look up the word *cycle*.

♦ What are the guide words on the page? How many meanings does the word *cycle* have?

♦ Find the word *cycle* in the vocabulary definitions.

♦ Which meaning from the dictionary best fits the word *cycle* as it is used in the vocabulary definitions? Why?

Infer Think about the life cycles of frogs or butterflies. What are the different stages that each animal goes through?

List the stages in the life cycle of a frog. List the stages in the life cycle of a butterfly next to your first list. Which animal goes through more phases? Why do you think frogs and butterflies lay so many eggs at one time?

 Skipping Difficult Questions Answer as many questions as you can in the time you have. Skip difficult questions and come back to them later, if you have time.

Multiple Choice Practice Read each question. Choose the best answer.

1 Which of the following things is an organism?

A

B

C

D

2 Which environment might be home to a polar bear?

A one with lots of sand
B one with lots of snow
C one deep in the ocean
D one high in a tree

3 What is one thing all plants need to survive?

A shelter
B sunlight
C oxygen to breathe
D food to eat

4 Where does a fish get its oxygen?

A It comes to the surface to breathe.
B A fish does not need oxygen.
C It gets oxygen from its food.
D It gets oxygen from the water.

Extended Response Practice Write your answer on a separate piece of paper.

5 Why do you think plants that live in water only grow near the surface? Think about what plants need to live.

Ecosystems

Getting Started · Quick Activity

- **<u>Discussion</u>** Water is a key part of this ecosystem. The fish needs water to live. The bear needs water too, but in a slightly different way! How do you use water?

- **<u>Critical Thinking</u>** What do you think would happen if all the water on Earth dried up?

23

Organisms Interact

▶ **Play Video: Introduction**

What happens in an ecosystem?

New Vocabulary

community all the living things in an ecosystem

ecosystem all the living and nonliving things in an environment and all their interactions

forest land area with a heavy growth of trees

habitat the home of a living thing

population all the members of a single type of an organism in an ecosystem

wetland land area that contains a lot of moisture

Soon You'll Know

Main Ideas

1. **What makes up an ecosystem**
2. **What makes up a community**
3. **Why a habitat is important to living things**

Remember The video you just watched described what's in an ecosystem. It's really not very hard to remember. That's because an ecosystem includes *everything* in an area and the interactions of those things!

Think about it One thing is for sure: there are lots of things in an **ecosystem.** An ecosystem includes all the living things in a certain area, such as the plants and animals. It also includes all the nonliving things, such as water and air.

Organisms live in ecosystems.

Ecosystems contain both living and nonliving things.

What goes on in an ecosystem is very important. Living things interact with each other. For example, many animals need to eat plants to live. They also use the nonliving things around them. Both plants and animals need air to survive. All these things existing together make an ecosystem.

Now You Know

Main Idea 1: Ecosystems

Living and nonliving things make up an ecosystem.

Remember You saw a wetland ecosystem in the video. How would you like to go fishing with the stork population in that community?

Think about it The living things in an ecosystem form a **community.** There are many different **populations** in a community. Populations include all the members of a single type of organism living in that area.

Here is a population of bison.

Different communities can have different populations.

For example, in a moist **wetland** area, you might find populations of ducks and cattails. In a heavily wooded area such as a **forest,** you might find a population of deer. The different types of trees also make up populations. Lots of different populations live together as part of a healthy community.

Now You Know

Main Idea 2: Communities

All living things in an ecosystem make up a community.

Remember In the video you learned that an organism's habitat is important. If you think about everything you get from your own home, you would probably agree!

Think about it An organism needs a lot to survive. Food, water, oxygen, and shelter are usually the basics. If they are going to do the organism any good, all these things have to be found where the organism lives.

A habitat must meet an organism's needs.

Sometimes habitats change, and organisms must move.

Luckily, that's exactly how it works. Organisms get what they need from their **habitats,** or homes. But habitats can change. For example, natural disasters such as fires and floods can change a habitat. People can change an organism's habitat too. When that happens, organisms sometimes have to move to new habitats.

Now You Know

Main Idea 3: Habitats

Living things get everything they need to live from their habitat.

Build Your Vocabulary

Vocabulary Review

Use the word bank to complete each statement.

1. A _____ is a land area with a heavy growth of trees.

2. An _____ includes all the living and nonliving things in an environment and all their interactions.

3. All the members of a single type of organism living in the same area are called a _____.

4. A _____ is an area that contains a lot of moisture.

5. All the living things in an ecosystem make up a _____.

6. An organism's home is called its _____.

community

ecosystem

forest

habitat

population

wetland

Word Study: In Your Own Words

1. Write the words below on one side of an index card.

| ecosystem | community | population |

2. On the back of each card, write a definition in your own words.

3. Write a sentence using each word. Create a drawing to go with each word as well.

Check Your Understanding

Show What You Know

Main Ideas: Write the answer to each question.

1. What does an ecosystem include?

2. What part of an ecosystem does the community include?

3. How does a habitat help an organism survive?

Critical Thinking

1. **Analyze** Explain how the destruction of a habitat can affect a whole ecosystem.

2. **Comprehend** Why are nonliving things so important in an ecosystem?

 Words **in Science**

Academic Vocabulary The word *nonliving* begins with the prefix *non-*. What does *nonliving* mean?

♦ The prefix *non-* means "not." Write how the word *nonliving* uses the meaning "not."

♦ What are some other words with the prefix *non-*? Look in a dictionary for help. How do these words mean "not"?

Process Skill **Quick Activity**

Communicate Think about the ecosystem on your school playground or a park. What are the living and nonliving parts? Can you identify different populations?

Make a poster that describes your playground ecosystem. List or draw all the different populations. Show nonliving things as well. List a few points to explain why nonliving things are important.

Energy in an Ecosystem

▶ **Play Video: Introduction**

How does energy move through an ecosystem?

Soon You'll Know

Main Ideas

1. How plants get and use energy
2. Why animals depend on plants
3. How energy moves from one organism to the next

Remember In the video you learned that every path of energy starts with the sun. No wonder that bright light in the sky is so important!

Think about it If you feel hungry, you might make yourself some food. But the way you make a peanut butter and jelly sandwich is quite different from the way a plant makes food! A plant depends on sunlight for energy to make its food.

The sun provides energy.   Plants use energy from the sun to make food.

The leaves and other green parts of a plant collect the sun's energy. Then they use that energy, along with air and water, to make sugars to use as food. Without the sun to start the process, the food-making activity could not happen.

Now You Know

Main Idea 1: Solar Energy

Plants use sunlight to produce their own food for energy.

Remember In the video you learned that the sun's energy can be passed from plants to animals. Sometimes this is easy to see. Other times it's not so easy.

Think about it As you know, a plant, such as grass, uses the sun's energy to make its own food. When a rabbit eats the grass, some of the energy that came from the sun is passed on to the rabbit.

Some animals, such as this rhinoceros and this butterfly, rely on plants for energy.

Some animals eat other animals.

If a hawk eats a rabbit, a little more of the sun's energy is passed on to the hawk. In this way, you can see how animals get energy directly from plants or from eating other animals that eat plants. Because the hawk eats the rabbit, it is called a **predator.** Because the rabbit is the one that gets eaten, it is the hawk's **prey.**

Now You Know

Main Idea 2: Predator and Prey

Animals get energy either from plants or from other animals that eat plants.

Remember In the video you learned about food chains and food webs. It's either eat or be eaten in the natural world!

Think about it There's a name for how energy is passed from one organism to the next. It's called a **food chain.** A bird eating a beetle that has eaten a leaf is a very simple food chain. Several food chains that link together in an ecosystem form a **food web.**

Plants use energy from the sun.　　**A caterpillar eats leaves on a plant.**　　**A bird eats the caterpillar.**

When organisms in the food web die, energy is still passed along. **Decomposers,** such as mushrooms or earthworms, break down dead plant and animal material. Important nutrients that plants need pass into the soil. This allows plants to begin their food-making process, and the cycle starts all over again.

Now You Know

Main Idea 3: Food Chains

Energy passes from one organism to the next through food chains.

33

Build Your Vocabulary

Vocabulary Review

Use the word bank to complete each statement.

1. An animal that hunts other animals for food is a _____.

2. A path that energy follows as it moves from one organism to another is a _____.

3. A _____ breaks down dead plant and animal material.

4. Many interlocking food chains come together to form a _____.

5. An animal that gets eaten by another animal is a _____.

decomposer

food chain

food web

predator

prey

Word Play: Puzzle

Think of a word that belongs with the rest of the words in each group. The dashes tell you how many letters are in the missing word or phrase.

1. lion, hawk, wolf, shark, _ _ _ _ _ _ _ _

2. grasshopper, mouse, sparrow, minnow, _ _ _ _

3. mushroom, bacteria, earthworm, mold, _ _ _ _ _ _ _ _ _ _

Check Your Understanding

Show What You Know

Main Ideas: **Write the answer to each question.**

1. What do plants need to produce their own food?

2. How do animals get energy?

3. What shows how energy is passed from one organism to the next in an ecosystem?

Critical Thinking

1. **Analyze** Explain why it is correct to say that all animals depend on plants.

2. **Apply** What might happen if one whole population of animals died out in an ecosystem?

Math in Science

Solve a problem Use math to solve the following ecology problem. Eagles are predators. Rabbits, mice, and fish are their prey.

- An eagle needs to eat 2 rabbits a day.
- In one area, there are 3 eagles.
- What is the minimum number of rabbits needed in the area for all the eagles to get what they need?

Process Skill Quick Activity

Infer Think of a forest ecosystem. Imagine all the living and nonliving things that are in that ecosystem. Now imagine that a bolt of lightning starts a fire that destroys much of the forest.

List things that might happen to the living things in the ecosystem. Then make another list, and try to infer what might happen to the nonliving things. Which part of the ecosystem do you think would be most affected by the fire?

DESERT ADAPTATIONS

Two things are for sure about life in the desert.
1. There are extreme temperatures.
2. There isn't very much water.

So how do plants and animals survive there? Special desert adaptations, of course!

BEAVERTAIL CACTUS

SAVING WATER The paddle-like stems of this cactus store water for the plant. A waxy covering helps seal in moisture. The roots of the beavertail cactus grow very close to the surface. That way they are ready to soak up every drop of rain.

DESERT STINKBUG

FIGHTING OFF ENEMIES When frightened, this beetle does a headstand by putting its front legs down and kicking its rear legs into the air. To drive a predator away, the stinkbug releases a very smelly substance. It usually works too! Who wants to eat a stinky dinner?

36

CONSERVING ENERGY The burrowing owl is really a borrowing owl! It lives in old burrows built by other animals. To stay cool, the owl stays inside during the hot day. Then at dusk the owl comes out to hunt.

BURROWING OWL

Critical Thinking

● Why do you think so many desert animals come out at night to hunt?

● Some desert plants grow extremely deep roots. Why do you think this might help those plants?

COOL BEHAVIOR This lizard knows how to stay cool. It has long legs and feet with long toes. So when the desert sand is hot, the collared lizard stands on tiptoes.

COLLARED LIZARD

Find Out More!

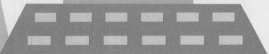

Research on the Web

Find out about other desert plants and animals. How do their special adaptations help them survive?

STORING FOOD Large kangaroo-like feet help this small rodent hop across desert sand. The kangaroo rat gets its water from the seeds it eats. It never has to take a drink!

KANGAROO RAT

Organisms Compete

▶ **Play Video: Introduction**

How do organisms compete to survive?

Soon You'll Know

Main Ideas

1. **Why plants and animals compete for resources**
2. **How plants survive in many places**
3. **What adaptations help animals to survive**

Remember In the video you learned that plants and animals compete for what they need. It's not quite the same as racing to be first in the lunch line, but you get the idea.

Think about it Rainfall doesn't come often to a hot, dry **desert.** And when it does come, it disappears quickly. Plants and animals that live in the desert often have to compete for what little water is available.

There isn't much water in the desert.

A rain forest has lots of plants and animals that compete for resources.

Rain forests have plenty of water. But the plants and animals that live there still have to compete. Plants need space and sunshine. Animals need food to eat and water to drink. When there are only limited amounts of **resources,** the organisms that compete the best will probably be the ones that survive.

Now You Know

Main Idea 1: Competition

Plants and animals compete with one another for resources like water and food to survive.

Remember The video you just watched was about plant adaptations. It's a good thing plants have special traits. After all, a tree can't put on a coat when it's cold!

Think about it You already know that plants have needs. You also know that plants must compete to meet these needs. What helps them do this? **Adaptations,** of course! Adaptations are special traits that help an organism survive.

Cactus needles are an adaptation.

Flowers attract insects that help flowering plants reproduce.

Remember the spines on the cactus in the video? They are an adaptation that helps defend the cactus. Water is scarce in the desert, so the cactus stores water in its stem. But thirsty animals often want to eat it. Spines are a plant adaptation that helps protect the cactus from these predators.

Now You Know

Main Idea 2: Plant Adaptations

Plants have many adaptations that help them survive in many places.

Remember The video you watched was about animal adaptations. Now you know why a giraffe is better suited to dine on the tops of trees than you are!

Think about it Some animal adaptations are easy to see. You can't miss the sharp claws on a bear. But other adaptations aren't so obvious. When a bear **hibernates** through the winter, it doesn't need to eat as much food. When caribou **migrate** to warmer places in winter, they are able to find the food they need to stay alive.

Animals have many
different adaptations.

Mimicry and camouflage are adaptations
that help some animals survive.

Remember the caterpillar with the large eye spots? It looks like a snake, so birds stay away. This **mimicry** adaptation helps the caterpillar survive. Other animals use **camouflage** to blend into their surroundings. This is another good adaptation that helps some animals avoid being seen.

Now You Know

Main Idea 3: Animal Adaptations

Animals have many adaptations, including camouflage, mimicry, and hibernation to name only a few.

Build Your Vocabulary

Vocabulary Review

Use the word bank to complete each statement.

1. A dry area that gets very little rainfall is known as a _____.

2. An adaptation that allows animals to blend into their surroundings is called _____.

3. An adaptation in which an animal imitates another animal is called _____.

4. An animal is said to _____ when it rests or sleeps through the cold winter.

5. Some animals move to another place, or _____ in search of food.

> camouflage
>
> desert
>
> hibernate
>
> migrate
>
> mimicry

Word Play: Idioms

Idioms are phrases that have special meanings.

food for thought

This phrase means an idea to think about, not food to eat.

hit the road

1. Explain what you think *hit the road* means.

2. What vocabulary word do you think best fits with *hit the road?* Explain why you think so.

Check Your Understanding

Critical Thinking

1. **Analyze** Explain why different animals have different types of adaptations.

2. **Apply** Can you think of some adaptations that help people survive?

Writing in Science

Write a paragraph Describe three different animal adaptations.

- Use at least three science vocabulary words.
- Put details in your paragraph.
- Remember, a good paragraph includes:
  - a topic sentence
  - a body
  - a closing sentence

Process Skill — Quick Activity

Observe Look at fish in an aquarium, or find some pictures of fish in a book or on the Internet. Observe the fish closely. Think about both how they look and how they act.

Now make a list. Describe all the adaptations you can see that help the fish survive. Are there any adaptations fish have that you can't see? List those adaptations too.

 TIPS

Reading Carefully Read each test question carefully. Read each answer choice and compare it to the question. Look for key words that will help you locate the correct answer.

Multiple Choice Practice Use the diagram to answer each question.

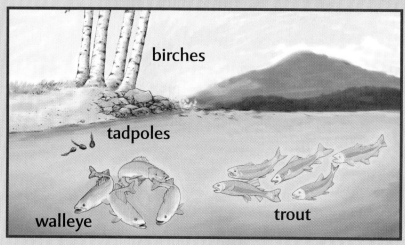

birches

tadpoles

walleye

trout

1 Which population could also live in this river community?

A camel population
B polar bear population
C whale population
D bass fish population

3 Which organism has the smallest population size in this community?

A tadpoles
B trout
C walleye
D birch trees

2 What is the wet land area around the river most likely to be called?

A desert
B forest
C wetland
D grassland

4 Which is part of the river community?

A birch trees
B rocks
C sunlight
D water

Extended Response Practice Write your answer on a separate piece of paper.

5 What will happen to the animals in the river when they die? How will this help the living plants?

 STOP

CHAPTER 3

Humans

Getting Started / Quick Activity

- **Discussion** Take a big bite out of life! Enjoying life and taking care of yourself are two really good ideas. How is the boy in this picture doing both at once? What do you do to stay healthy?

- **Critical Thinking** How do you think staying healthy helps you stay happy?

People Have Needs

▶ **Play Video: Introduction**

What do people need to stay healthy?

New Vocabulary

carbohydrates nutrients found in food from plants that give your body energy

minerals nutrients that work to help control body processes

nutrients substances that provide energy and control body processes

protein a substance found in plants and animals that keeps a body strong

vitamin a nutrient in foods that the body needs for growth

Soon You'll Know

Main Ideas

1. What human beings need in order to survive
2. What humans need to do to stay healthy
3. What is in food that our bodies need

Remember In the video you learned about human needs. Taking good care of yourself is a full-time job, and you're the one in charge of it.

Think about it There are many choices to make every day. What should we eat? What should we drink? Should we walk or ride a bike? The list goes on and on. Sometimes it's easy to forget that the choices we make every day can affect how we feel.

People have many needs.   Exercise is important.   Water helps keep people healthy.

People need to breathe clean air that contains oxygen. We also need to eat healthful foods and drink clean water. A safe place to live is important too.

We make decisions each day that affect our health. So it's good to be making the right choices!

Now You Know

Main Idea 1: Human Needs

Humans need air, food, water, and shelter.

Remember You know about basic human needs. But in the video, you learned more about sleep, exercise, and diet. It turns out that it takes a little effort to keep your body running smoothly!

Think about it Have you ever noticed what happens to machines that are never used? Their batteries run down. They get squeaky. Sometimes they even start to rust. Our bodies are kind of like machines. They need some attention to run well.

Get enough sleep!   **Don't forget to exercise!**   **Be sure to eat right!**

Sleeping never sounds fun. But it sure is important to give your body a good rest. Of course, after you rest, you need to get things moving again with some exercise. And what **fuels** it all? A healthful diet, of course! You can't get very far on a diet of junk food.

Now You Know

Main Idea 2: Sleep and Exercise

People need to be physically fit, eat a balanced diet, and get plenty of rest to be healthy.

▶ **Play Video C: Nutrition**

Remember In the video you learned about the nutrients your body needs. They're like a grocery list for the body that you have to shop for every day.

Think about it It sounds pretty simple. You need to eat plenty of **nutrients** to get energy and to control your body's functions. But you can't go to the store and buy one box of all the nutrients you need. You have to get nutrients from lots of different sources.

Labels show nutrition facts.   Different foods have different nutrients.

Carbohydrates come from plants. You need to eat grains, fruits, and vegetables to get them. **Protein** comes from both plants and animals. If you want to build strong muscles, you can eat things like meat, eggs, nuts, and beans. **Vitamins** are also found in the foods that come from plants and animals. **Minerals,** such as the calcium found in milk and cheese, keep your bones and teeth strong and healthy.

Now You Know

Main Idea 3: Nutrition

Foods contain many nutrients that are important to the body.

Build Your Vocabulary

Vocabulary Review

Use the word bank to complete each statement.

1. All the substances that our bodies need every day to have energy and control body processes are called _____.

2. _____ are nutrients found only in plants.

3. Meat, eggs, and beans are good sources of _____, which helps build strong muscles.

4. Calcium is an example of a _____ that helps build strong bones and healthy teeth.

5. A _____ is a nutrient that our bodies need for growth.

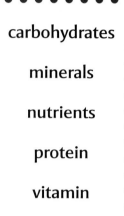

carbohydrates

minerals

nutrients

protein

vitamin

Word Study: Classifying

Draw three squares and label each as shown below.

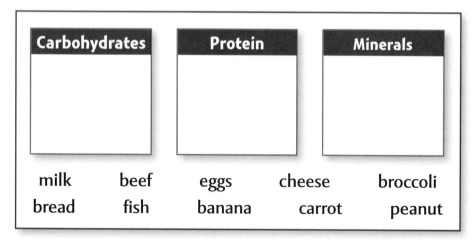

Carbohydrates	Protein	Minerals

milk   beef   eggs   cheese   broccoli

bread   fish   banana   carrot   peanut

1. Look at the foods above.

2. Decide if each food is a source of carbohydrates, protein, minerals, or more than one nutrient.

3. Write the food in the correct square on your paper.

Check Your Understanding

Show What You Know

Main Ideas: Write the answer to each question.

1. What are four basic needs of all humans?

2. What are three different things people can do to stay healthy?

3. What kinds of nutrients does your body need each day and where can you get them?

Critical Thinking

1. **Comprehend** Why do people need to eat a variety of foods?

2. **Evaluate** Explain why physical education classes in school can be important.

 Words **in Science**

Academic Vocabulary Using context clues, write what you think each boldfaced word means and why you think so.

♦ We tried to find the **source** of the strange noise.

♦ Building a computer is a long and difficult **process.**

♦ Not getting enough sleep seemed to **affect** his ability to play.

Process Skill **Quick Activity**

Classify Now you know how important diet and exercise are to your body. Think about the choices you make each day to be healthy.

Make two columns on a piece of paper. Label one list "Diet" and the other "Exercise." For two days, write what you eat and how you exercise. When you are finished, decide if you made healthful choices.

Move It!

Exercise sometimes sounds like work. But it's really just play. From kickball to riding a bike, any movement counts as exercise.

Jump!

DID YOU KNOW?

In an average lifetime, a human heart will beat over two and a half billion times!

Sweat!

Your steady heartbeat keeps you alive. But like all the rest of the muscles in your body, your heart needs a workout to stay healthy.

Some kinds of exercise build strength in your muscles. Strong muscles mean more power!

Push Yourself!

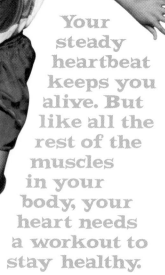

Bend!

Not everyone can be as flexible as this. But you should be able to touch your toes! Bending and stretching help keep your body flexible and pain free.

Dance

Martial Arts

Yoga

Try these activities to stretch you out!

CHALLENGE ZONE

Critical Thinking

● Why do you think it's important to do different kinds of exercise?

● Why do you think even kids need to exercise?

Find Out More!

Research on the Web

Find out more about the exercise that helps your heart. It's called aerobic exercise. What kind of exercises are aerobic?

Microorganisms and People

▶ **Play Video: Introduction**

How do microorganisms affect people?

New Vocabulary

bacteria one-celled living things

disease illness or sickness of a plant or animal

fungus a one- or many-celled organism that absorbs food from other organisms

microorganism an organism that is so small you need a microscope to see it

microscope a device that uses glass lenses to allow people to see very small things

mold a type of fungus that grows in damp places

Soon You'll Know

Main Ideas

1. The various sizes of living things
2. How microorganisms can help humans
3. How microorganisms can be harmful to humans

▶ **Play Video A: Magnification**

Remember In the video you learned about microscopes and microorganisms. You can now think of the word *tiny* in a whole new way!

Think about it Sometimes it's hard to see a small mosquito buzzing around your head. But imagine trying to look at a **microorganism** that is hundreds of times smaller than the period at the end of this sentence. Impossible? Not with the special view that a **microscope** can give you.

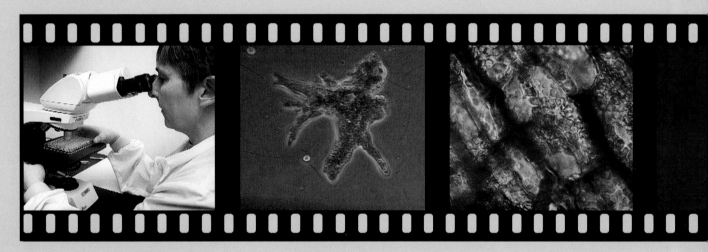

Microscopes are useful tools.　　　Tiny microorganisms and cells cannot be seen without a microscope.

Microscopes can magnify, or enlarge, even the smallest microorganisms. Tiny, one-celled **bacteria** suddenly come into view through the lens of a microscope. Without a microscope, you would never know that these little organisms exist!

Now You Know

Main Idea 1: Magnification

Microorganisms are so small, they can only be seen using a microscope.

55

Play Video B: Helpful Microorganisms

Remember The video showed you how some microorganisms are helpful. It's hard to believe that such tiny organisms can have such a big effect on people!

Think about it Bacteria are sometimes the superheroes of the microscopic world. From cleaning up oil spills to making yummy yogurt, bacteria are often very helpful to humans.

Most **decomposers** are microorganisms. They break down dead material and add nutrients to the soil that help plants grow.

Bacteria can be used to make yogurt.

You need a microscope to see some fungi, but others are easy to see.

Now it's time to talk about the **fungus** among us! Some types of fungi are very small. For example, a small fungus, known as a **mold,** can be used to make helpful medicines, such as penicillin. But there are other types of fungi that aren't so small. A mushroom, which is a larger fungus, can be a tasty topping on your favorite pizza!

Now You Know

Main Idea 2: Helpful Microorganisms

Some microorganisms, like the mold used for penicillin, are helpful to people.

▶ **Play Video C: Harmful Microorganisms**

Remember The video showed you how harmful some small microorganisms can be. There just had to be a bad side to these tiny creatures, didn't there?

Think about it Although some microorganisms are helpful, some of them can cause problems. Certain types of bacteria can cause **disease,** or illness, in humans. Strep throat is an illness that is caused by bacteria.

Microorganisms are making this child sick.

Plants, such as these potatoes, can be harmed by microorganisms too.

Humans aren't the only targets of harmful microorganisms. Plants and animals can get sick because of microorganisms. Even property can be affected. Remember that fungus called mold? Sometimes whole houses have to be torn down after mold starts growing in damp walls.

Now You Know

Main Idea 3: Harmful Microorganisms

Microorganisms can cause disease in humans, plants, and animals.

Build Your Vocabulary

Vocabulary Review

Use the word bank to complete each statement.

1. An illness in a plant or animal is called a _____.

2. A _____ is a device that allows us to see very small things.

3. One type of fungus that grows in damp places is a _____.

4. A _____ can be a one-celled or a many-celled organism.

5. An organism that is too small to be seen with the human eye is called a _____.

6. _____ are always one-celled organisms.

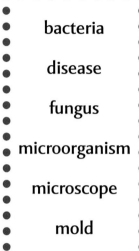

bacteria

disease

fungus

microorganism

microscope

mold

Word Study: Synonyms

Synonyms are words with the same meanings.

Match the words in column A with their synonyms in column B.

A	B
1. bacteria	a. illness
2. disease	b. magnifying tool
3. fungus	c. mushroom
4. microscope	d. single-celled organism
5. microorganism	e. tiny living thing

Check Your Understanding

Main Ideas: Write the answer to each question.

1. How big are microorganisms, and how can they be seen?

2. How are microorganisms helpful to people?

3. How can microorganisms be harmful to people?

Critical Thinking

1. **Analyze** Explain why scientists want to get rid of some microorganisms but not others.

2. **Synthesize** Why is it a good idea to cover your nose and mouth when you sneeze?

Math in Science

Measure Magnification can make things look 1,000 times bigger. What would you look like if you were 1,000 times bigger?

♦ How much would you weigh?
♦ How tall would you be?
♦ How long would your foot be?
♦ How long would your arm be?

Process Skill — Quick Activity

Infer When astronauts first returned to Earth after exploring the moon, they were kept away from other people for a while. Scientists watched them to make sure they didn't get sick. Luckily, everyone turned out to be fine.

Explain why you think scientists decided they had to watch the astronauts for a while.

People and the Ecosystem

▶ **Play Video: Introduction**

How do ecosystems change?

New Vocabulary

conserve to save, protect, or use something wisely without wasting it

endangered close to becoming extinct; having very few of its kind left

extinct died out; as when a species no longer exists

pollution the adding of harmful substances to the water, air, or land

resource a material on Earth that is necessary or useful to living things

Soon You'll Know

Main Ideas

1. How humans can change an ecosystem
2. How ecosystems can change naturally
3. How humans can help to conserve and protect ecosystems

Remember In the video you learned how humans can affect an ecosystem. Because we are all part of an ecosystem, our actions will have some sort of effect on the ecosystem. Sometimes it's not in a good way!

Think about it Imagine you are enjoying a warm, sunny day in the park. You drop a piece of paper, but you don't stop to pick it up. After all, one little piece of paper isn't going to ruin that big park! Is it?

People use natural resources.

Pollution affects the land.

Pollution affects the air.

Of course, a single piece of paper won't ruin the park. But if everyone kept dropping things in the park, they would soon build up and start to ruin that natural **resource.** People have to take good care of the environment. **Pollution** can ruin our valuable water, air, and land.

Now You Know

Main Idea 1: Pollution

Human actions, such as pollution, can damage an ecosystem.

▶ Play Video B: Changing Ecosystems

Remember In the video you learned that natural disasters can occur. Some things that affect the environment are out of our control.

Think about it You might have seen floods, fires, and even erupting volcanoes on the news. Natural disasters like these happen from time to time. All of these events change the ecosystems in which they occur.

Hurricanes can cause floods.   Fires burn parts of the ecosystem.   Ecosystems can recover.

Ecosystems have an amazing way of bouncing back! For example, after a forest fire, it won't be long before tiny shoots of new growth start to spring up through the ashes.

Ecosystems try to stay in balance. It's the natural way of things.

Now You Know

Main Idea 2: Changing Ecosystems

Natural events such as fires, floods, and volcanoes can change ecosystems.

Remember In the video you learned how people can protect ecosystems. What a great feeling it is to do something good for Earth!

Think about it You might have heard about a bird called the dodo. But you've certainly never seen one. That's because dodos are **extinct.** This kind of bird no longer exists on Earth. It became extinct over one hundred years ago.

People can protect and conserve environments.

Some animals are endangered.

Other plants and animals are **endangered,** which means they are close to becoming extinct. So it's up to us to try to protect these living things and the places in which they live. Trying to protect habitats and **conserve** resources keeps ecosystems in better shape for all of us.

Now You Know

Main Idea 3: Conservation

The conservation of resources helps to maintain and protect ecosystems.

Build Your Vocabulary

Vocabulary Review

Use the word bank to complete each statement.

1. An _____ animal species is one that is close to dying out.

2. Animal species that are _____ no longer exist on Earth.

3. When we turn off lights, we _____ electricity.

4. A natural _____ is a material on Earth that is useful to living things.

5. Heavy traffic and smoke from factories can often cause air _____.

conserve

endangered

extinct

pollution

resource

Word Study: Antonyms

Antonyms are words with opposite meanings.

Match each word in Column A with its antonym in Column B.

A	B
1. conserve	a. overpopulated
2. polluting	b. alive today
3. resource	c. waste
4. endangered	d. cleaning up
5. extinct	e. useless object

Check Your Understanding

Main Ideas: Write the answer to each question.

1. How can human actions damage an ecosystem?

2. What are three different natural events can cause changes to an ecosystem?

3. What can people do to protect ecosystems?

Critical Thinking

1. **Analyze** Explain why it would not be a good thing if too many species of plants and animals died out in one area.

2. **Synthesize** Explain what might happen if everyone decided it was okay for them to litter.

Writing in Science

Write a paragraph Describe some of the things children can do to conserve and protect the environment.

- ◆ Use at least three science vocabulary words.
- ◆ Put details in your paragraph.
- ◆ Remember, a good paragraph includes:
  - a topic sentence
  - a body
  - a closing sentence

Process Skill — Quick Activity

Measure Take a look around your lunchroom at school. Watch how much trash is thrown away.

List each day of the week on a piece of paper. Estimate how much trash is thrown away each day from the lunchroom. Add up the numbers for a whole week. Can you think of any way that your school could reduce the amount of trash it produces?

Sequence A sequence is a list based on an order. When you are asked to put items in order, make sure you know whether the sequence is based on time, location, or size.

Multiple Choice Practice Read each question and look at the pictures. Choose the best answer.

A animal is gone **B** natural disaster **C** loss of habitat **D** animal in habitat

1 Which order of pictures shows the correct sequence of events?

A picture A, B, C, then D
B picture D, B, C, then A
C picture C, B, A, then D
D picture D, C, A, then B

3 When a disaster occurs, which picture shows what happens next?

A picture A
B picture B
C picture C
D picture D

2 Which picture shows what happened last?

A picture A
B picture B
C picture C
D picture D

4 Which type of natural disaster could replace the picture of the fire and still result in the same sequence?

A earthquake
B volcanic eruption
C flood
D all of the above

Extended Response Practice Write your answer on a separate piece of paper.

5 Think of five organisms and put them in order from largest to smallest. Make sure to include a microorganism in your list.

CHAPTER 4

Reshaping Earth

Getting Started — Quick Activity

- **Discussion** Surf's up! It may look like a surfer's dream, but this wavelike structure is solid rock. Wind and water have carved the rock into this interesting shape. How long do you think it took for this to happen?

- **Critical Thinking** Why do you think Earth's surface looks so different in so many places?

Forces Shape the Land

▶ **Play Video: Introduction**

How does Earth's surface change?

New Vocabulary

earthquake a sudden movement in the rocks that make up Earth's crust

erosion the carrying away of weathered materials

glacier a large mass of moving ice

landform a natural feature on Earth's surface

volcano an opening in the surface of Earth from which lava flows

weathering the process that causes rocks to crumble, crack, and break

Soon You'll Know

Main Ideas

1. How to describe different parts of Earth's surface
2. Why some changes to Earth's surface happen slowly over time
3. Why some changes to Earth's surface happen quickly

Remember Mountains, rivers, and valleys are just some of the features that dot Earth's surface. No wonder it's taken thousands of years for explorers to map this **planet!**

Think about it What does the land look like where you live? Is it flat or hilly? Do you live near water? What about mountains—are there any in view? Does the slope of the land change dramatically near your town?

Rivers are bodies of water that can shape the land.

Beaches are found along the coasts.

Valleys shelter cities and towns.

It's amazing how many different **landforms,** or natural features, there are on Earth's surface. These landforms have a big effect on the people who live near them. For example, you might go fishing on a lake. You might ski down a mountain. As you can guess, life in a dry **desert** is very different from life in a steamy, wet rain forest.

Now You Know

Main Idea 1: Landforms

Earth's surface can be described by its many different landforms and bodies of water.

Remember In the video you learned that even **solid** rock can be broken down. But don't hold your breath—it can take a very long time for that to happen!

Think about it Waves crash against a rocky shore. Month after month, year after year, the water pounds the rocks. You may not see changes, but they are happening. Finally, the rocks begin to crack, crumble, and break. **Weathering** is taking place. Large rocks are being broken into smaller pieces.

The surface of Earth can change.

Water and gravity can cause erosion.

Glaciers can move everything in their path.

What happens to the rocks when they break apart? They are carried away by **erosion.** Different forces on Earth can cause erosion.

Wind, water, gravity, and even huge pieces of ice, called **glaciers,** can cause the erosion of different landforms. Year after year, the surface of Earth is slowly changed by these forces.

Now You Know

Main Idea 2: Weathering and Erosion

Some changes to Earth's surface caused by wind, water, and ice happen slowly.

▶ Play Video C: Volcanoes and Earthquakes

Remember In the video you learned about changes that can happen very quickly on Earth. Sometimes the surface of this planet can really rock!

Think about it A **volcano** might sit quietly for years and years. But when it starts to erupt, everything suddenly changes! Burning clouds of hot **gas** and tons of volcanic ash might be thrown up into the atmosphere. Lava might flow out of the volcano, burning and burying everything in its path.

Hot lava rolls down the side of a volcano.

Earthquakes can happen without warning and cause lots of damage.

Earthquakes can also shake things up. They happen when pressure builds up beneath Earth's surface. This pressure causes big masses of rock to start sliding past each other. Sometimes earthquakes only cause a small rumble. But a strong earthquake can damage whole cities in minutes.

Now You Know

Main Idea 3: Volcanoes and Earthquakes

Volcanoes and earthquakes can cause rapid change to Earth's surface.

Vocabulary Review

Use the word bank to complete each statement.

1. An opening in the surface of Earth from which lava flows is called a _____.

2. A sudden movement in the rocks that make up Earth's crust can cause an _____.

3. _____ is the process that causes rocks to crumble, crack, and break.

4. A natural feature, such as a river or mountain, that is found on Earth's surface is a _____.

5. Wind and water can carry away material in a process called _____.

earthquake

erosion

glacier

landform

volcano

weathering

Word Play: Rhyming

Poems and songs often use words that rhyme, or that have the same sounds at the end. For example:

I heard a distant rumble and the ground began to shake.

In a second it was over—it had been a small earthquake!

Now you do it. Write your own poem using one or more of the new vocabulary words. Make sure you include words that rhyme.

Check Your Understanding

Show What You Know

Main Ideas: Write the answer to each question.

1. What kinds of words can be used to describe Earth's surface?

2. What processes slowly change Earth's surface?

3. What events can change Earth's surface more quickly?

Critical Thinking

1. **Analyze** Explain why you think scientists want to develop earlier warning systems for earthquakes.

2. **Apply** Tell why you think plants are sometimes grown on steep roadside cliffs.

 Math in Science

Calculate California has many strong earthquakes each year. In one year, there were 162 earthquakes in California. In that same year, there were 7 earthquakes in Oregon and 24 earthquakes in Washington. What was the total number of earthquakes on the West Coast that year?

Process Skill Quick Activity

Predict The Richter scale is used to describe the force of an earthquake. Find a copy of the Richter scale. Describe what effects an earthquake measuring 5 on the Richter scale might have on your classroom and the students in it.

73

Materials on Earth

▶ **Play Video: Introduction**

How do rocks and soil form?

New Vocabulary

fossil the preserved remains or trace of an organism that lived long ago

humus decayed plant or animal material in soil

mineral nonliving material that can be found in rocks and soil

soil a mix of tiny rock particles, minerals, and decayed plant and animal materials

Soon You'll Know

Main Ideas

1. What rocks are made of and how they form
2. What we can learn from fossils
3. What soil is made of and why it's important

Remember In the video you learned about rocks. Well, guess what? If you have any interest in becoming a collector, you can become a rock hound!

Think about it Most rocks may seem pretty dull. But you would probably change your mind if you could see all the **minerals** inside them! Scientists have identified more than 4,000 minerals in the rocks of Earth's crust. But only about 100 of them are common.

Rocks are all around us.   Rocks are made of minerals.   Rocks form in three different ways.

Rocks form in three different ways. Some rocks are made when hot **liquid** rises from deep within Earth and cools. Other rocks are formed when bits of rock and sand pile up and are pressed together. And other rocks change with intense heat and pressure deep inside Earth. It takes a long time, but new rocks are always forming!

Now You Know

Main Idea 1: Rocks and Minerals

Rocks are made of minerals and can be formed three different ways.

Remember Some rocks can show us clues about life from long ago. But it takes a lot of detective work to figure out what these clues mean!

Think about it Sometimes a rock holds more than just minerals. It may also contain a **fossil,** which is the preserved remains or traces of an organism that lived long ago! Scientists who study fossils can learn about extinct organisms and discover clues to Earth's environment in the past.

Some rocks contain fossils.

There are many different kinds of fossils. Some are more complete than others.

It's hard to get the whole story of the past from fossils. That's because complete organisms are usually hard to find. Many fossils provide just a piece of information. It's like a piece of a puzzle. Scientists must try to put these puzzle pieces together to form a more complete picture of the past.

Now You Know

Main Idea 2: Fossils

Fossils give us clues about plants and animals that lived long ago.

Remember In the video you learned about the importance of soil. Even though we may spend a lot of time washing off dirt, it's really something that we can't live without!

Think about it **Soil** is an important **natural resource.** It's made of tiny rock particles, minerals, and decayed plant and animal materials. There are many different kinds of soil. Some soil is **sandy,** while other soil is mostly **clay.**

Soil contains tiny rock particles broken down by natural forces.   **Worms and other animals create space for air in soil.**   **Soil captures water.**

A lot of plants grow well in soil that contains **humus,** which is decayed material that is rich in **nutrients.** Plants provide most of the material in humus, while **insects,** worms, and other animals mix the humus into the soil. With a steady supply of **air** and water, plant roots take what they need from the soil. Then the plants will grow strong and healthy.

Now You Know

Main Idea 3: Soil

Soil is a mix of rock, humus, air, and water and is very important to plants and animals.

77

Build Your Vocabulary

Vocabulary Review

Use the word bank to complete each statement.

1. Decayed plant or animal material mixed with tiny rock particles is called _____.

2. Every rock is made up of at least one _____.

3. The preserved remains or trace of an organism that lived long ago is a _____.

4. The decayed plant or animal material in soil is called _____.

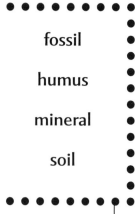

fossil

humus

mineral

soil

Word Study: In Your Own Words

1. Write the words below on one side of an index card.

mineral	humus	fossil

2. On the back of each card, write a definition in your own words. Add a drawing that explains the word.

3. Write a sentence using each word.

Check Your Understanding

Main Ideas: **Write the answer to each question.**

1. What are rocks made of, and how do they form?

2. What can we learn from fossils?

3. What is soil, and why is it important?

Critical Thinking

1. **Evaluate** Why do you think plants grow well in some soils but not in others?

2. **Analyze** Why do you think fossils are so hard to find?

Writing in Science

Write a letter Pretend you are a dinosaur hunter. Write a letter to a friend to tell him or her about your latest fossil find.

Remember to include these parts in your friendly letter:

♦ Begin with a greeting.

♦ Indent each paragraph in the letter.

♦ End with a closing, followed by your name.

Process Skill — Quick Activity

Observe Look around the area near your school. What kind of soil is there? Are there lots of trees and plants? Is most of the area paved?

Make a list of what you think you would need to grow tomatoes and carrots near your school. What kind of soil is in the area? Is it sandy, or is it mostly clay? Would you need to grow the vegetables in pots of soil? If so, what kind of soil do you think would be best?

A VERY RARE FIND

Sue
the *T-Rex*

Sue is the largest and most complete *Tyrannosaurus rex* dinosaur ever found.

DID YOU KNOW?

Sue's skeleton is over 90% complete. That means most of the bones in her body were found together. A very rare find indeed!

The Discovery

On the morning of August 12, 1990, a fossil-hunting team's truck had a flat tire near Faith, South Dakota. Some of the team went into town to get the tire fixed. But fossil hunter Sue Hendrickson stayed behind. Good thing she did! She found a few fossils that were part of the largest complete *T. rex* skeleton ever found!

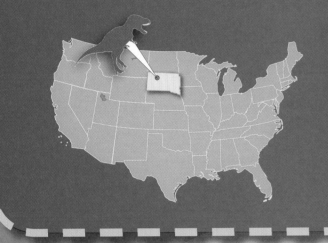

Sue's Stats

Scientific name:
Tyrannosaurus rex
(lizard king)

Age:
About 67 million years

Height (at hips):
4 meters

Weight (estimated):
6.4 metric tons

Favorite food:
meat

Sex:
Unknown*

* This *T Rex* is named "Sue" for the fossil hunter who discovered her, Sue Hendrickson. But scientists don't really know if this dinosaur was male or female.

Sue's fossilized skull weighs 272 kg. It's so heavy that it can't be displayed with the rest of her bones.

Earth's Resources

▶ **Play Video: Introduction**

How do we use natural resources?

New Vocabulary

fossil fuel a substance, such as coal or oil, that was formed millions of years ago from the remains of plants and animals

inexhaustible resource one that cannot be used up easily

natural resource a material on Earth that is necessary or useful to people

nonrenewable resource one that cannot be reused or replaced easily

renewable resource one that can be replaced in a short period of time

Soon You'll Know

Main Ideas

1. How we use Earth's many resources to make and grow things
2. How Earth's resources are used for fuel
3. Why we need to conserve resources

Remember Earth gives us a lot. Water, air, land, and trees all come from Earth. It's nature's grocery store for our every need!

Think about it Long ago, people used to live completely off the land. That means they grew all of their own food using the soil around their homes. They also built shelters using only the materials around them. Earth was the main supplier of any **natural resource** they needed.

Soil and water are natural resources.

Minerals, such as copper and lead, are used to make many things.

Of course things have changed for a lot of people. New technology allows people to make new things. But we still depend on Earth's resources for many of our needs. Although manufacturing and processing can produce new and different materials, everything still comes from Earth. Using natural resources never goes out of style!

Now You Know

Main Idea 1: Natural Resources

Whether we're building homes, making clothing, or growing food, we use Earth's resources for many things.

Remember Coal, oil, and natural gas may look pretty boring. But the history of how these energy sources formed is very interesting!

Think about it Even though it's weird to imagine, tiny ancient sea creatures helped make the fuel that moves our cars! That's because gasoline is made from oil. Oil is a **fossil fuel** that was formed millions of years ago, partly from the remains of ocean creatures that were smaller than the period at the end of this sentence.

Oil is made into gasoline which fuels our cars.

Coal was formed millions of years ago.

Burning fossil fuels can cause pollution.

Coal is another fossil fuel that we burn mostly to make electricity. Like oil, coal was formed millions of years ago and is a good source of energy. But there is a downside to burning coal and oil. These fuels can also cause air pollution.

Now You Know

Main Idea 2: Fossil Fuels

Burning fossil fuels provides needed energy, but also produces unwanted pollution.

Remember You hear it all the time. Turn off the lights! Save water! Don't litter! But now you know why protecting our resources is such a good idea!

Think about it Rivers and lakes may run low in certain seasons. But, chances are, rain will eventually fill these bodies of water back to normal levels. That's because water is a **renewable resource,** which means it can usually be replaced in a short period of time.

Water is a renewable resource.

Fossil fuels are nonrenewable.

Conservation saves energy and the environment.

The same is not true of fossil fuels. Because they take millions of years to form, once these **nonrenewable resources** are used up, they cannot be easily replaced. Although the sun and wind are **inexhaustible resources** that cannot easily be used up, most other resources need to be protected and preserved so they last longer.

Now You Know

Main Idea 3: Conservation

Conservation saves Earth's resources and improves our quality of life.

Build Your Vocabulary

Vocabulary Review

Use the word bank to complete each statement.

1. Because the sun always shines, it is a good example of an _____.

2. Fossil fuels are examples of _____.

3. A _____ includes any natural material on Earth that is necessary or useful for people and other organisms to survive.

4. Coal, natural gas, and oil are _____ that were formed millions of years ago.

5. Because trees are a source of fuel that can grow back in a fairly short time, they are considered a _____.

> fossil fuels
>
> inexhaustible resource
>
> natural resource
>
> nonrenewable resources
>
> renewable resource

Word Study: Examples

Match each vocabulary word in column A with the best example in column B.

A	B
1. inexhaustible resource	a. coal
2. fossil fuel	b. sun
3. renewable resource	c. water

Check Your Understanding

Main Ideas: **Write the answer to each question.**

1. How do we use some of Earth's resources?

2. Why is burning fossil fuels both a good and bad thing to do?

3. How can people conserve energy, and why is that important?

Critical Thinking

1. **Analyze** Explain why scientists are always looking for new sources of energy.

2. **Synthesize** Why do you think it might be a good idea to find ways to use the sun and wind for energy?

Writing in Science

Quick write Take a stand.

There is only so much money available for energy development. So where do you think we should spend most of the money? Should people look for new oil deposits or develop new sources of energy?

Pick which position you favor. Then give facts to support it.

Process Skill — Quick Activity

Interpret data Think about how you use energy each day. Do you leave the water running when you brush your teeth? Do you turn off lights when you leave a room?

Make a T-chart. List what you do every day to save energy on one side of the chart. List what you do every day that wastes energy on the other side. What are your conclusions?

 TIPS

Understanding Maps Review the map quickly. Refer back to it as you answer the questions. A key or legend may provide helpful information.

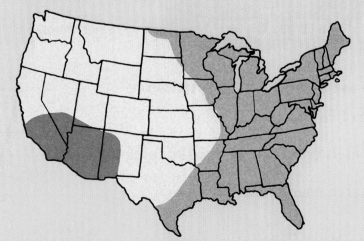

Average Daily Solar Radiation

June

KWh/m2/day

▨	4
☐	5
▨	7

Multiple Choice Practice Read each question. Choose the best answer.

1 What part of the year does the data in this map show?

A. the whole year
B. the summer
C. the month of June
D. You cannot tell from this map.

3 What part of the country gets the most solar radiation in January?

A. the yellow zone
B. the red zone
C. the green zone
D. You cannot tell from this map.

2 What part of the country gets the most solar radiation in June?

A. the red zone
B. the green zone
C. the yellow zone
D. the yellow and red zones

4 How much solar radiation does the green zone get each day in June?

A. 2 kWh/m2/day
B. 4 kWh/m2/day
C. 6 kWh/m2/day
D. 8 kWh/m2/day

Extended Response Practice Write your answer on a separate piece of paper.

5 You want to know if you can use solar power to heat your house in the winter. Would this map help you decide? Explain your answer.

Earth's Weather

Getting Started ⟩ Quick Activity

- **Discussion** Think of the bird's-eye view you'd have from these hot-air balloons. How high up do you think they can go? What do you think the weather might be like up there?

- **Critical Thinking** Many people take oxygen tanks with them in their balloons. Why is that a good idea?

Air Around You

▶ **Play Video: Introduction**

What happens in Earth's atmosphere?

New Vocabulary

air mass a large region of the atmosphere where the air has similar characteristics

air pressure the force put on a given area by the weight of the air above it

atmosphere the gases, or air, that surround Earth

front the boundary between air masses that have different temperatures

weather the condition of the atmosphere at a given time and place

wind moving air

Soon You'll Know

Main Ideas

1. What the air that surrounds Earth is called

2. What characteristics describe weather

3. Why the weather changes constantly

Remember In the video you learned about the atmosphere surrounding Earth. You can think of the atmosphere as a giant security blanket of air that protects us.

Think about it When astronauts go outside the International Space Station, they wear spacesuits and helmets and carry their own air supply. Why? They are at the outer limits of Earth's **atmosphere** where there is not enough oxygen to breathe.

The atmosphere supports life.

There is less oxygen in the layers of the atmosphere farthest away from Earth.

The atmosphere that surrounds Earth contains gases, such as nitrogen and oxygen. These are needed for life. There are several layers in the atmosphere, and each one is a little different. The layer of the atmosphere closest to Earth contains the most oxygen. It is also the layer where clouds, wind, and rain are found.

Now You Know

Main Idea 1: Atmosphere

Earth is surrounded by an atmosphere of gases, including the oxygen we breathe.

Remember In the video you learned about characteristics of weather. There are lots of ways to describe and measure what's going on outside!

Think about it How's the **weather** today? You may answer that it's hot or cold outside. That's because you know that air temperature is one characteristic of weather. But it's not the only one. Whether it is clear or cloudy, raining or snowing, you can check out the weather every day!

The sun heats Earth.   **Low air pressure can mean a storm.**   **Moving air is wind.**

The weight of the air pressing down all around us is **air pressure.** Air pressure is different all over Earth. Places with high air pressure tend to have sunnier weather, while places with low air pressure tend to have cloudy or even stormy weather. Why? Because air moves from high pressure areas to low pressure areas. And when the air moves, you get **wind.**

Now You Know

Main Idea 2: Weather

Air temperature, air pressure, and wind are characteristics of weather.

Remember In the video you learned how weather can change. It's all about the battle of the air masses!

Think about it When a warm **air mass** settles over your town, you are probably in for some warm and sunny days. An air mass is a large region of the atmosphere where the air has similar characteristics. If the wind kicks up, it may mean that a cold air mass is approaching, and that could mean big changes!

Warm Air   Cold Air

Air masses are
constantly moving.

When a cold front meets a warm front,
the weather will change.

The boundary, or **front,** between two air masses is where the biggest weather changes occur. A front is the boundary between air masses that have different temperatures. When a cold air mass collides with a warm air mass, strong thunderstorms may result. The weather changes all the time because air is always on the move.

Now You Know

Main Idea 3: Air Masses

The air on Earth constantly moves in the form of wind, which causes changes in the weather.

Build Your Vocabulary

Vocabulary Review

Use the word bank to complete each statement.

1. The _____ report lets us know the condition of the atmosphere at any given time and place.

2. Moving air is called _____.

3. The boundary between air masses that have different temperatures is called a _____.

4. The mixture of gases that surround Earth is called the _____.

5. _____ is the force put on a given area by the weight of the air above it.

6. An _____ is a large area of the atmosphere in which the air has similar characteristics.

air mass

air pressure

atmosphere

front

weather

wind

Word Study: In Your Own Words

1. Write the words below on one side of an index card.

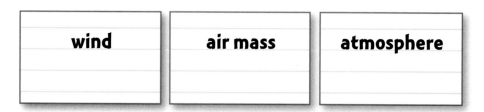

wind	air mass	atmosphere

2. On the back of each card, write a definition in your own words. Add a drawing that explains the word.

3. Write a sentence using each word.

Check Your Understanding

Show What You Know

Main Ideas: Write the answer to each question.

1. Why is the atmosphere around Earth important?

2. What are three important characteristics of weather?

3. What causes weather to change?

Critical Thinking

1. **Synthesize** What do you think would happen if our atmosphere suddenly disappeared?

2. **Apply** Explain why it is often windy before a storm.

 Writing in Science

Write a paragraph Describe a severe thunderstorm.

- ◆ Use at least three science vocabulary words.
- ◆ Try to think of how a storm affects all of your senses. For example:
  - Tell what the wind feels like.
  - Describe how the rain sounds.
  - Explain what you see in the sky.

Process Skill Quick Activity

Predict Try to watch the weather tonight on the evening news. Look at the weather map. Can you see the areas of high pressure and low pressure? What weather is happening in those different pressure areas?

Be a meteorologist: Predict where the weather will be changing tomorrow. How do you know?

HURRICANE HUNTERS

Most people try to get away from dangerous hurricanes. But not hurricane hunters! These brave men and women fly right into the eye of the storm. They know that the information they collect saves lives!

A WILD RIDE

Flying into the eye of a hurricane is wilder than a roller coaster ride. The crew straps in tight, but they still get tossed around. While inside the eye, planes can drop as much as 300 meters.

DROP THE DROPSONDE!

A dropsonde is a small tube with weather instruments in it that is connected to a parachute. When hurricane hunters reach the eye of a hurricane, they release the dropsonde from the plane. As the dropsonde drops through the hurricane, it sends back information on temperature, humidity, pressure, and wind speeds inside the huge storm.

Square-cone Parachute

Vents

Computer

GPS Receiver

Pressure sensor

Radio Transmitter

Humidity sensors and temperature sensor

GPS Antenna

Battery

DID YOU KNOW?

Hurricane-hunter planes carry enough fuel to fly for about fourteen hours. The average mission lasts about eleven hours.

Critical Thinking

● Why do you think it's so important for hurricane hunters to collect wind speed data?

● Why do hurricane hunters always fly out over the ocean?

Find Out More!

Research on the Web

Learn about the planes that fly into hurricanes. What instruments do they carry? Why are they able to withstand such strong winds?

Water Cycle

▶ **Play Video: Introduction**

How does water move on Earth?

New Vocabulary

condensation when water cools and changes from a gas into a liquid

evaporation when water warms and changes from a liquid into a gas

precipitation water in the atmosphere that falls to Earth as rain, snow, hail, or sleet

water cycle the constant movement of water from Earth to the atmosphere and back to Earth again

water vapor water in the form of a gas

Soon You'll Know

Main Ideas

1. Where water is found on Earth
2. How water moves in a constant cycle
3. Why it's important to protect and conserve water

Remember In the video you learned about sources of water. We don't call Earth "the blue planet" for nothing, you know!

Think about it Where do you get your water? If you say you get it from the sink, you are only partly right. Where did the water come from before it came to your sink? What's the *real* source of the water? It could have come from a river or a lake or from deep underground. It might even have been frozen in a glacier at one time.

| Water covers most of Earth. | Rivers, lakes, and streams are sources of freshwater. | Glaciers are frozen sources of freshwater. |

Water covers three-fourths of Earth's surface. Much of this is salt water in the ocean. But there are plenty of freshwater sources too. You know about freshwater sources such as lakes and rivers. You might even have a pond near you. But did you know that glaciers are a big source of freshwater as well?

Now You Know

Main Idea 1: Sources of Water

Most of Earth's water is found in oceans, rivers, lakes, and glaciers.

▶ **Play Video B: Water Cycle**

Remember Water isn't the kind of resource that gets used up. It's something that gets used over and over again.

Think about it The sun provides the energy for the water cycle. The sun warms water, which then **evaporates** into a gas called **water vapor.** As water vapor rises into the air, it cools, and **condensation** can happen. This means the water vapor changes back into a liquid. So what happens next?

The water cycle keeps water moving on Earth.

Water vapor condenses to form clouds.

Precipitation is part of the water cycle.

These tiny drops of water form clouds in the air. But when these water droplets get too heavy, they fall back to Earth. This is the **precipitation** we all know as rain, hail, or sleet. In this way, the **water cycle** constantly moves water from Earth to the atmosphere and back again.

Now You Know

Main Idea 2: Water Cycle

Evaporation, condensation, and precipitation are part of the water cycle.

Remember In the video you learned why we need to protect our water sources. It's a no-brainer! Who wants to drink polluted water?

Think about it Do you remember the last time you were thirsty? Really thirsty? A nice, cool drink of water probably would have tasted really good then. You may not think about it often, but water is one of our most precious **resources.**

Animals need clean water.   Plants need water to grow.   Polluted water can move through the water cycle.

What do we have to do to protect our water supply? The main job is to keep it clean. But it's not a simple task. As you've seen, **pollution** can move through the water cycle. This means pollution of the air and land can affect our water later on. It is important for each one of us to do what he or she can to keep water safe and clean.

Now You Know

Main Idea 3: Conserving Water

All living things need water. Water can become polluted and unsafe for use.

Build Your Vocabulary

Vocabulary Review

Use the word bank to complete each statement.

1. The constant movement of water from Earth to the atmosphere and back to Earth again is called the _____.

2. Rain, snow, hail, and sleet are known as _____.

3. _____ occurs when water warms and changes from a liquid into a gas.

4. _____ is water in the form of a gas.

5. _____ occurs when water cools and changes from a gas into a liquid.

condensation

evaporation

precipitation

water cycle

water vapor

Word Study: Classifying

Draw two circles and label each circle as shown below.

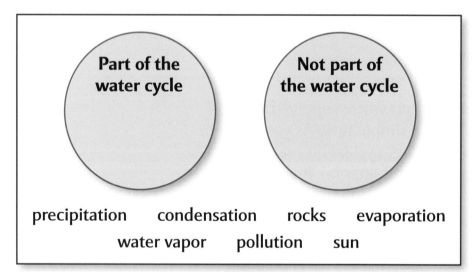

Part of the water cycle

Not part of the water cycle

precipitation   condensation   rocks   evaporation
water vapor   pollution   sun

1. Read each term in the word bank.

2. Decide whether it is part of the water cycle or not.

3. Write the term in the correct circle on your paper.

Check Your Understanding

Main Ideas: Write the answer to each question.

1. Where is water found on Earth?

2. How does the water cycle move water around Earth?

3. Why is it important to keep water clean?

Critical Thinking

1. **Analyze** Discuss how air pollution affects the water cycle.

2. **Evaluate** Explain why it's not okay to pollute even faraway areas of water.

Math in Science

Compare The chart below shows normal rainfall for Boston and actual rainfall for Boston in 2006.

	Normal	Actual in 2006
March	10 cm	less than 1 cm
April	9 cm	4.5 cm
May	8 cm	32 cm

Which month received the most actual rainfall? Which month normally receives the most rainfall?

Process Skill — Quick Activity

Interpret data Keep a record of the weather over the next week. Record whether it is sunny or cloudy. Check your local newspaper to find out how much precipitation falls each day.

At the end of the week, look at your data. Do you think the temperature and precipitation levels were high, low, or average?

Where could you get more information to tell whether you are right?

103

Predicting Weather

▶ **Play Video: Introduction**

How do scientists predict weather?

New Vocabulary

anemometer a tool that measures wind speed

barometer a tool that measures air pressure

hurricane a huge storm that forms over warm ocean water

meteorologist a person who studies the weather

thermometer a tool that measures temperature

tornado a violent spinning wind that moves across the ground in a narrow path

Soon You'll Know

Main Ideas

1. How meteorologists gather data to describe weather

2. About different types of severe weather

3. Why it's important to stay alert to weather forecasts

Remember In the video you learned about some of the tools used to measure weather. It takes a lot more than a temperature reading to report the weather these days!

Think about it People who study the weather, called **meteorologists,** use lots of tools. A **thermometer** is a tool that measures temperature. A **barometer** keeps track of the rise and fall of air pressure. The spin of an **anemometer** tells how fast the wind is blowing.

Meteorologists study
the weather.

Meteorologists use many tools to
gather data about the weather.

Those tools are not very high-tech. However, that's not the whole story. Meteorologists use satellites and computers too. The data these machines analyze give meteorologists a more complete weather picture than they've ever had before.

Now You Know

Main Idea 1: Weather Instruments

Meteorologists use tools to measure and describe weather.

Remember In the video you learned about severe weather. Sometimes Mother Nature can dish up some nasty storms!

Think about it Every June, all eyes turn toward the southern and eastern coasts of the United States as **hurricane** season begins. Hurricanes can form any time. But it is warm ocean water that fuels these huge storms. Large hurricanes can kill many people and destroy many buildings.

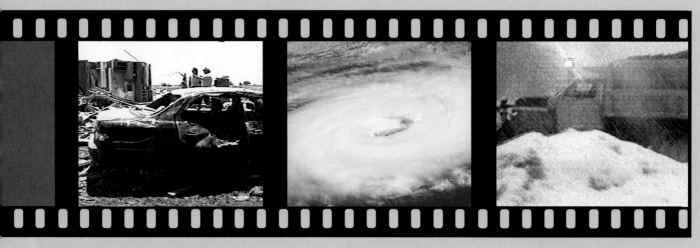

Tornadoes can cause much damage.

Hurricanes are huge storms.

Blizzards can be hazardous.

Tornadoes are another threat. These violent windstorms form over land instead of water. They are most common in places where the land is very flat and open. When it gets colder, there is always the threat of blizzards. A blizzard can make travel extremely difficult. The cold, snow, and wind in these storms combine to cause trouble.

Now You Know

Main Idea 2: Severe Weather

Storms such as blizzards, tornadoes, and hurricanes can cause severe damage.

Remember What do you do when bad storms approach? A little preparation and a lot of common sense can go a long way to keep you safe!

Think about it Bad weather happens no matter where you live. Because you can't avoid it, you have to learn to prepare for it. Pay attention to the weather and take precautions when necessary.

Weather forecasts help people prepare.

Safety drills are important.

BE PREPARED
Emergency Plan
Batteries
Radio and Flashlight
Bottled Water
Non-Perishable Food
Emergency Supplies

Prepare ahead of time for emergencies.

Even when the weather is not severe, it's good to know what to expect so you can plan your day. But when there is a tornado, good planning can save your life! You must get to a safe spot very quickly. Hurricanes move more slowly, so there is time to gather emergency supplies or to leave ahead of the storm. But no matter what, good information and planning can help keep people safe.

Now You Know

Main Idea 3: Staying Safe

People can take action to stay safe in any kind of weather.

Build Your Vocabulary

Vocabulary Review

Use the word bank to complete each statement.

1. A _____ is a person who studies the weather.

2. A tool that measures temperature is called a _____.

3. A _____ is a huge storm that forms over warm ocean water.

4. An _____ measures wind speed.

5. A tool that measures air pressure is called a _____.

6. A _____ is a violent, spinning wind that moves across the ground in a narrow path.

anemometer

barometer

hurricane

meteorologist

thermometer

tornado

Word Study: Word Roots

Many science words come from other languages. Those parts are called roots.

The root *therm* means "heat" in Greek. The vocabulary word *thermometer* uses this root.

| thermo | + | meter | = | ***thermometer*** |

1. Look in your dictionary for three more words that use the root *therm*.

2. Write the words and their definitions.

3. Write a sentence using each word.

Check Your Understanding

Show What You Know

Main Ideas: **Write the answer to each question.**

1. What tools do meteorologists use to measure and describe weather?

2. What effects can severe storms have on people and property?

3. What can people do to stay safe in severe weather?

Critical Thinking

1. **Analyze** Explain why meteorologists try to collect so much weather information.

2. **Evaluate** Summarize your school's safety drills.

Writing in Science

Write a letter to the editor Explain why people should have an emergency plan in case of severe weather.

♦ Give reasons why people should have a plan.

♦ Try to make your letter convincing.

♦ At the end of your letter, restate the purpose of the letter.

Process Skill — Quick Activity

Draw conclusions Use the Internet to find out how many hurricanes were predicted for last year's Atlantic hurricane season. Then find out how many actual hurricanes occurred. Are the numbers very different? Why do you think it might be hard to predict hurricanes?

Using Graphs When you read a graph, you have to match the information shown at the left of the graph with the information shown at the bottom of the graph.

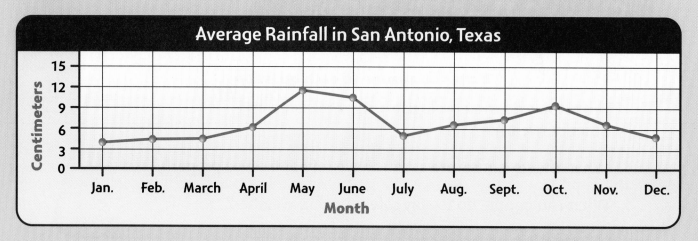

Average Rainfall in San Antonio, Texas

Centimeters / *Month*

Jan. Feb. March April May June July Aug. Sept. Oct. Nov. Dec.

Multiple Choice Practice Use the graph to answer the questions below.

1 How much rain falls in July?

A 5 cm
B 13 cm
C 8 cm
D none

2 Which months have about the same amount of rainfall as August?

A July and September
B September and October
C April and November
D none

3 Which month has the most rain?

A January
B May
C October
D December

4 Which month has the least rain?

A January
B February
C July
D December

Extended Response Practice Write your answer on a separate piece of paper.

5 Do you think you would need your umbrella more often in April or in June? Explain your answer.

STOP

Space

Getting Started / Quick Activity

- **Discussion** Some planets are much bigger than Earth, and some are much smaller. You could fit 750 Earths inside Saturn. Why do you think Saturn looks like a star when it is viewed without a telescope?

- **Critical Thinking** What do you think Saturn's rings are made of?

Earth, Sun, and Moon

▶ **Play Video: Introduction**

How do the sun and the moon affect Earth?

New Vocabulary

axis an imaginary line through the center of Earth

gravity a pulling force between two objects, such as Earth and you

phase the shape of the lighted part of the moon

revolve to move in a nearly circular path around something else

rotate to spin on an axis

seasons spring, summer, fall, and winter

Soon You'll Know

Main Ideas

1. What causes day and night on Earth
2. Why the seasons change
3. Why the moon looks different throughout the month

Remember In the video you learned about Earth's spin. It's not something that will make you dizzy. But the spin does have something to do with making you sleepy!

Think about it The **force** of attraction between two objects, such as you and Earth, is called **gravity.** Because Earth has more **mass** than you, Earth's gravity keeps you in place. This is a good thing because even though you can't feel it, Earth is always spinning.

| Earth rotates in space. | Facing the sun brings daylight. | Rotating away from the sun brings nighttime. |

But what does this spinning mean for us on Earth? When Earth **rotates,** or spins, on its **axis,** the side of Earth facing the sun is in daylight. As Earth continues to rotate, this side of Earth eventually turns away from the sun. Then daytime turns into nighttime. And nighttime is time to sleep!

Now You Know

Main Idea 1: Day and Night
Earth's rotation causes day and night.

Remember In the video you learned that it takes one year for Earth to make one trip around the sun. And while that's happening, we get to enjoy the seasons!

Think about it As you know, the **seasons** include spring, summer, fall, and winter. In some places, the temperature changes a lot between the seasons. In other places, the weather doesn't change very much at all.

| Earth revolves around the sun. | Winter days can be cold and short. | Summer days can be longer and hotter. |

Seasons occur because of Earth's tilt on its axis during its **revolution** around the sun. As Earth slowly revolves, the part of Earth that is tilted farthest away from the sun gets less light. So that area has winter. The part of Earth that is tilted toward the sun gets the sun's rays more directly. Longer and warmer days make it summer for those places.

Now You Know

Main Idea 2: Seasons

The tilt of Earth on its axis causes the seasons.

▶ **Play Video C: Phases of the Moon**

Remember In the video you learned about phases of the moon. It's pretty neat to think that because the moon can't make its own light it reflects light from the sun.

Think about it Poets have written great things about the full moon, but the sun should really get the credit. After all, it is light from the sun that reflects off the moon so we can see it in our nighttime sky.

The sun's light reflects off the moon.

The amount of reflected sunlight we see on the moon determines which phase it is.

As the moon moves around Earth, we see different **phases** of the moon. During a new moon, the side of the moon that is facing away from Earth is lit by the sun. As the moon revolves around Earth, more of the lighted side comes into view, until we have a full moon. During a full moon, the entire sunlit side of the moon is visible from Earth. After the full moon, you see less of the moon's sunlit side each night until we have a new moon again.

Now You Know

Main Idea 3: Phases of the Moon

The moon appears to change shape during phases.

Build Your Vocabulary

Vocabulary Review

Use the word bank to complete each statement.

1. _____ is the force that keeps us all on Earth.

2. The shape of the lighted part of the moon is called a _____ of the moon.

3. The imaginary line through the center of Earth is its _____.

4. To _____ is to spin on an axis.

5. To _____ means to move in a nearly circular path around something else.

6. Winter, spring, summer, and fall are known as _____.

axis

gravity

phase

revolve

rotate

seasons

Word Play: Rhyming

Poems and songs often use words that rhyme, or have the same sounds at the end. For example:

> Earth's rotation
> May spin our nation
> But it's not the reason
> For our seasons

Now you do it. Write your own poem using one or more of the new vocabulary words. Make sure you include words that rhyme.

Check Your Understanding

Show What You Know

Main Ideas: Write the answer to each question.

1. What causes day and night on Earth?

2. What causes seasons on Earth?

3. Why does the moon appear to change shape during its different phases?

Critical Thinking

1. **Analyze** Explain why our seasons happen in one year.

2. **Synthesize** What do you think would happen to one side of Earth if Earth did not rotate as it moved around the sun?

 Words **in Science**

Academic Vocabulary Use the dictionary to look up the word *revolve*.

♦ How many meanings does the word *revolve* have?

♦ Find the word *revolve* in the vocabulary definitions.

♦ Which meaning from the dictionary best fits the word *revolve* as it is used in the definitions? Why?

Process Skill **Quick Activity**

Infer Think about how Earth's movement causes day and night.

Now imagine that you put a wet beach towel in the sun to dry. Two hours later the towel is in the shade.

Did the towel move? Did something else move? Explain what happened.

The Sun and Planets

▶ **Play Video: Introduction**

What makes up our solar system?

New Vocabulary

orbit the path an object follows as it revolves around another object

planet a large, round, or nearly round body that revolves around the sun and has cleared a path within its own orbit

solar system includes the sun and all the objects that orbit the sun

star a huge, hot sphere of gases that gives off its own light

Soon You'll Know

Main Ideas

1. What the sun and planets are like and what they do

2. Which four planets are closest to the sun

3. Which four planets are farthest from the sun

Remember In the video you learned that the sun is the star of our solar system. Though we can see many stars in the night sky, it's the sun that gives us light and heat.

Think about it The sun, like every other **star**, is a huge, hot ball of gases that gives off its own light. It's also the center of our **solar system.** The solar system includes the sun and all other objects, such as planets and moons, that **orbit** the sun.

The sun is a star.　　　Eight planets orbit the sun, including Venus and Jupiter.

Our planet Earth is one of eight large bodies, or **planets,** that orbit the sun. Earth is the third closest to the sun. You may be able to remember the names of all eight planets by using the sentence you learned in the video: **M**y **V**ery **E**nergetic **M**other **J**ust **S**erved **U**s **N**achos. Remember, the letter at the beginning of each word is the first letter of a planet. If we discover any more planets, we'll have to think of a new sentence!

Now You Know

Main Idea 1: Solar System
The sun is a star orbited by planets.

Remember In the video you learned about the inner planets of our solar system. Now you know why Earth is sometimes called "the third rock from the sun"!

Think about it Earth is one of four planets that are closest to the sun. Mercury is closest to the sun, followed by Venus, Earth, and Mars. Because these planets are closest to the sun, it makes sense that they are the warmest. But their temperatures can vary a lot.

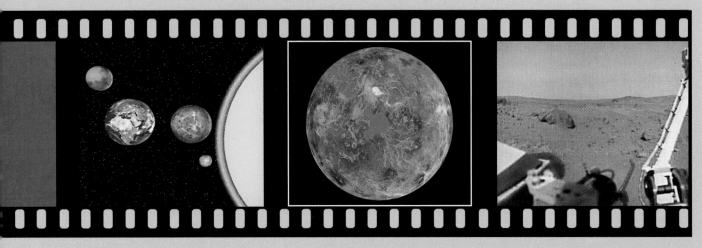

The four inner planets are closest to the sun.

Venus and Mars are very different from Earth.

Mercury has no atmosphere at all, while Venus has thick clouds that trap in the heat from the sun. Mars has a very thin atmosphere. All of these planets are small and rocky like Earth, but Earth is the only planet with liquid water on its surface and oxygen in its atmosphere. As far as we know, Earth is also the only planet that supports life. But scientists are still looking!

Now You Know

Main Idea 2: Inner Planets

Mercury, Venus, Earth, and Mars are the inner planets of our solar system, closest to the sun.

Remember The outer planets of the solar system really rock—but all of them are made of **gases**! And they are really, really big too.

Think about it The four outer planets of the solar system include Jupiter, Saturn, Uranus, and Neptune. These planets are huge and mostly made of gases. That's why they're sometime called the "gas giants." Jupiter is the biggest of them all. Saturn has several broad rings. Uranus and Neptune are similar in size and color.

Jupiter is the
largest planet.

Saturn is known
for its rings.

Pluto is a dwarf planet.

Pluto, which used to be grouped with the outer planets, now belongs to a third category, the **dwarf planets.** Because Pluto is not very large, and it shares its orbit with other objects, astronomers decided that Pluto belonged in this third group. The large asteroid Ceres and a newly discovered object called Eris are also dwarf planets.

Now You Know

Main Idea 3: Outer Planets

Jupiter, Saturn, Uranus, and Neptune are the outer planets, farthest from the sun.

Build Your Vocabulary

Vocabulary Review

Use the word bank to complete each statement.

1. The sun is a _____ , which is a huge, hot ball of gases that gives off its own light.

2. Earth is a _____, one of eight large bodies that orbit the sun.

3. The _____ includes the sun and all the objects that orbit the sun.

4. The path an object follows as it moves around another object is an _____ .

> orbit
>
> planet
>
> solar system
>
> star

Word Play: Idioms

Idioms are phrases that have special meanings.

food for thought

This means "an idea to think about," not "food to eat."

1. Explain why you think Earth is sometimes called "the blue planet."

Hint: think about what Earth looks like from space and why.

2. Explain why you think Earth is sometimes called the "third rock from the sun."

Check Your Understanding

Show What You Know

Main Ideas: Write the answer to each question.

1. What is the sun made of, and where is it located?

2. What are the four inner planets in our solar system?

3. What are the four outer planets?

Critical Thinking

1. **Analyze** Explain why discovering water on another planet would be an important find.

2. **Evaluate** Why do you think it is so difficult to explore other planets?

Writing in Science

Write a comparison Describe what you think the planet Mars is like. Compare and contrast it to Earth.

♦ Write about its distance from the sun, its temperature, and its weather.

♦ Tell why you think it might be a good or bad place to live.

♦ Research Mars to check your ideas.

Process Skill Quick Activity

Communicate Pretend you are making a time capsule that will be sent into outer space. Your task is to describe Earth.

Describe five things about Earth that make it a good place to live. Think of what an alien from another planet might like to know!

An Eye in the Sky

Large telescopes on Earth can give a closeup view of stars and planets millions of kilometers away. But did you know there is another telescope that has an even better view of space? It's called the Hubble Space Telescope. It looks into space from 600 kilometers above Earth!

Ready to Retire?

The Hubble Space Telescope was launched into space on April 24, 1990. Since then it has taken about 750,000 pictures. Without getting some kind of space "tune-up," it will probably stop taking pictures sometime in 2008.

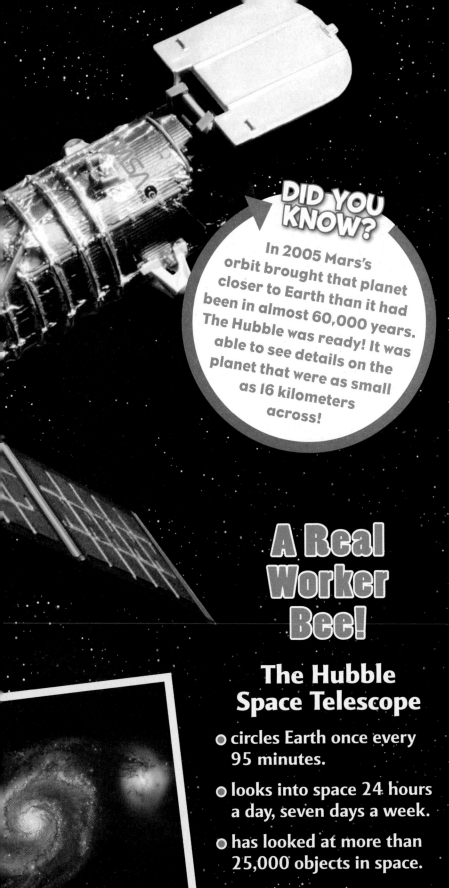

DID YOU KNOW?

In 2005 Mars's orbit brought that planet closer to Earth than it had been in almost 60,000 years. The Hubble was ready! It was able to see details on the planet that were as small as 16 kilometers across!

A Real Worker Bee!

The Hubble Space Telescope

- circles Earth once every 95 minutes.
- looks into space 24 hours a day, seven days a week.
- has looked at more than 25,000 objects in space.

Critical Thinking

- Why do you think it's a good idea to send a telescope into space to take pictures?

- Do you think that the Hubble should be repaired or replaced? Why?

Find Out More!

Research on the Web

The Hubble is not alone. Go to www.nasa.gov and enter "space telescopes" in the search box to find out about new space telescopes and their missions.

Exploring Space

▶ **Play Video: Introduction**

How do scientists study space?

New Vocabulary

constellation a group of stars that appears to form a pattern

rocket a tube-shaped device driven through the air or space by a stream of hot gases

satellite any object that orbits another larger body in space

space probes rocket-launched vehicles that carry data-gathering equipment into space

telescope a tool that gathers light to make faraway objects appear closer

Soon You'll Know

Main Ideas

1. How early people studied the night sky
2. How telescopes can be used to study space
3. Other ways that scientists study space

▶ **Play Video A: Constellations**

Remember In the video you learned how ancient people used the stars as their guide. It helped them to keep track of the changing seasons too.

Think about it It may be hard for you to see the imaginary picture in some **constellations.** You may wonder how in the world anyone could have thought a group of stars formed that pattern. But the picture is not the important part. Watching a group of stars to find out about changing seasons is what matters!

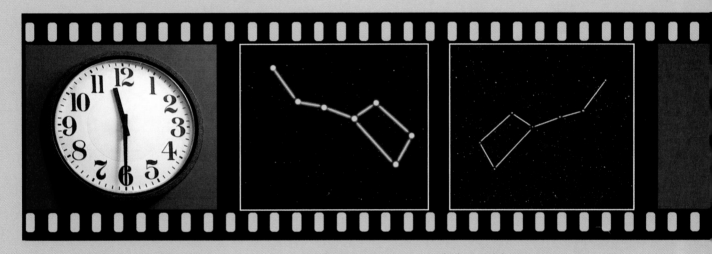

Early people used the stars to keep track of time.

The Big Dipper and the Little Dipper are constellations you might know.

Constellations appear to be moving across the sky because Earth rotates on its axis. The part of the sky we see each night changes as Earth revolves around the sun. This means we see different constellations in different seasons. One constellation that you might know is the Big Dipper. The Big Dipper is part of another constellation called *Ursa Major,* which means "the Great Bear."

Now You Know

Main Idea 1: Constellations

Early people named patterns in the night sky. These patterns are called constellations.

127

Remember In the video you learned how we need a little help to see into outer space. One tool in particular has brought outer space into sharper view!

Think about it Can you find the moon in the night sky? No problem! Your eye can easily see an object that close. You might even be able to see a bright star. But wait a minute—are you sure it's a star? It could be a planet. It's difficult to tell without a little help.

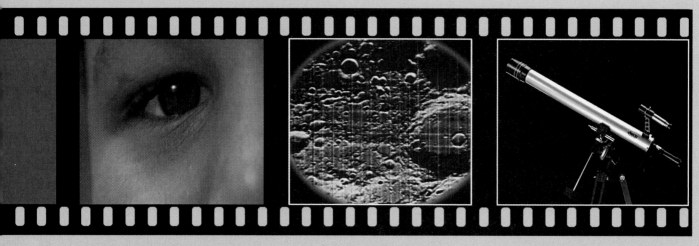

The human eye can't see very far into space.

Galileo saw craters on the moon.

Modern telescopes show us a whole new world.

Telescopes are tools that have helped people study the night sky for centuries. Most telescopes use a curved piece of glass called a lens to gather light and make an object look larger and closer. Large, modern telescopes let scientists on Earth view faraway stars, planets, and galaxies in greater detail.

Now You Know

Main Idea 2: Telescopes

Telescopes are helpful tools used to study space from Earth.

▶ Play Video C: Rockets

Remember In the video you learned about the many ways we blast off this planet to explore outer space. There is still much more to explore!

Think about it Have you ever heard someone say that it doesn't take a rocket scientist to figure something out? Well, it *does* take a rocket scientist to figure out space! Or, at least how to get there. Because of Earth's gravitational attraction, we need **rockets** to lift instruments and people into space.

Rockets are used to carry people into space.   **Space probes can visit other planets.**   **Satellites help cell phones work.**

Some rockets launch into space with astronauts on board. Other rockets don't carry people. Instead, they send **space probes** to faraway planets to gather information. Rockets send **satellites** into space all the time. These objects orbit Earth and allow us to send and receive information all around the planet. Some satellites gather information to help meteorologists predict the weather.

Now You Know

Main Idea 3: Rockets

The space shuttle, space probes, and satellites are some tools that allow scientists to study outer space.

Build Your Vocabulary

Vocabulary Review

Use the word bank to complete each statement.

1. Any object that orbits another larger body is called a _____.

2. The tube-shaped device that was used to send astronauts to the moon is called a _____.

3. A group of stars that appears to form a pattern is called a _____.

4. A _____ is a rocket-launched vehicle that carries data-gathering equipment into space.

5. A tool that gathers light to make faraway objects appear closer is a _____.

constellation

rocket

satellite

space probe

telescope

Word Study: Examples

Match each vocabulary word in column A with the correct example in column B.

A	B
1. constellation	a. Hubble
2. telescope	b. the Big Dipper
3. space probe	c. space shuttle
4. satellite	d. Spirit
5. rocket	e. moon

Check Your Understanding

Show What You Know

Main Ideas: **Write the answer to each question.**

1. What did ancient people use to keep track of the seasons?

2. What is a common tool used to study space from Earth?

3. What are three ways that scientists gather information from space?

Critical Thinking

1. **Analyze** Explain why ancient people thought the constellations actually moved across the sky.

2. **Evaluate** Why do you think scientists send space probes instead of rockets with astronauts to planets like Mars?

Writing in Science

Quick Write Choose one of the topics below and write what you think.

♦ What was the most interesting part in the video on exploring space?

♦ What would it be like to launch into space on a rocket?

♦ Would you like to be an astronaut some day? Explain why or why not.

Process Skill — Quick Activity

Observe Look at the sky tonight and see what you can see. Can you see stars? Can you see the moon? Look in the same direction at the same time over the next three nights. Record what you see.

Does anything change in the night sky? What do you think you could see if you had a telescope to view the same scene?

Using Charts Charts often use numbers to show information. Read the question first. Then look in each column for the answer. Don't be distracted by the extra information on the chart.

Multiple Choice Practice Use the chart to help you answer each question.

Inner Planet	Time to Orbit the Sun (in Earth days)	Number of Moons
Mercury	88	0
Venus	224	0
Earth	365	1
Mars	687	2

1 Which planet takes longer to orbit the sun than Earth does?

A Mercury
B Venus
C Mars
D Earth takes the longest.

3 How many days longer does Earth take to orbit the sun than Venus?

A about 500 days
B about 140 days
C 1 less
D 1 more

2 Which planet orbits the sun faster than Venus does?

A Mercury
B Venus is the fastest
C Mars
D Jupiter

4 Which planet has the most moons?

A Mercury
B Venus
C Earth
D Mars

Extended Response Practice Write your answer on a separate piece of paper.

5 The word *year* means "the time it takes to orbit the sun." If you are 8 Earth years old, about how old are you in Venus years? Explain your answer.

STOP

CHAPTER 7

Force and Motion

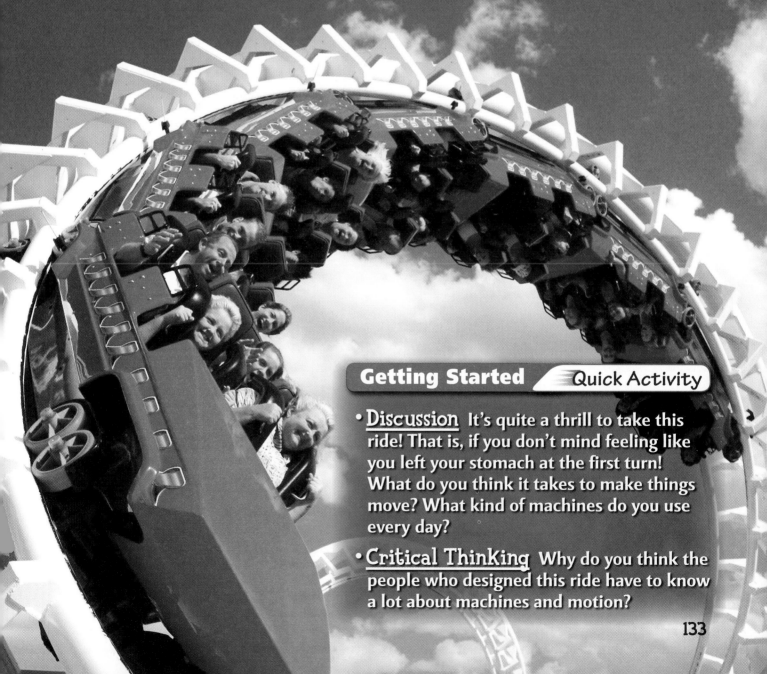

Getting Started / Quick Activity

- **Discussion** It's quite a thrill to take this ride! That is, if you don't mind feeling like you left your stomach at the first turn! What do you think it takes to make things move? What kind of machines do you use every day?

- **Critical Thinking** Why do you think the people who designed this ride have to know a lot about machines and motion?

133

Forces Make Things Move

▶ **Play Video: Introduction**

What causes objects to move?

Soon You'll Know

Main Ideas

1. **What causes an object to move or what changes an object's motion**

2. **How friction affects the way an object moves**

3. **The effect of gravity on motion and on us**

Remember In the video you learned about pushes and pulls. Our world is filled with constant tugs-of-war!

Think about it Do you remember the last time you waited in line? You were in one **position,** or location, and you didn't move until a **force** put you in **motion.** That force probably came from your own muscles and bones. But it could have been a friendly push from the person behind you too!

Motion can be fun!   Unbalanced forces make things move.

Every force is either a push or a pull. And forces work in pairs. When pairs of forces are balanced, there is no motion. But when forces are unbalanced, you'd better stand back or hold on. That's because motion is sure to happen. Pull on that rope hard enough, and the other team will come tumbling over.

Now You Know

Main Idea 1: Force

Objects can only move if they are pushed or pulled.

Remember In the video you learned about how some forces can slow you down. You might even say that friction can be a real drag!

Think about it It seems to make sense, doesn't it? If it takes a force to get things moving, then it will probably take a force to make things stop. One of those forces is friction. **Friction** occurs when one object rubs against another. It's a force that slows things down.

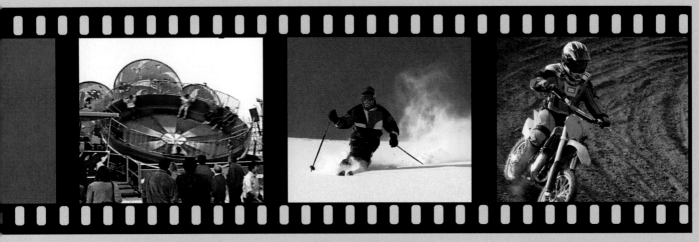

| Friction will help stop this ride. | Less friction helps things move faster. | More friction slows movement down. |

Sometimes friction is a good thing. It can keep a car's tires from sliding off the road. It can also stop you from sliding on ice. But other times, friction can be a problem. It's not good when friction wears down the gears in machinery. Friction can also heat things up when you don't want to feel the heat.

Now You Know

Main Idea 2: Friction

Friction is a force that slows things down.

▶ **Play Video C: Gravity**

Remember In the video you saw astronauts floating around in space. You even saw food floating. Can you imagine having to chase your food around to eat it?

Think about it What happens when you stumble and trip? You fall down, of course. But did you ever wonder why you don't fall up? It's because there is another force, **gravity,** that exists between you and Earth. Gravity is always at work.

Gravity brings us back down.   **There is no gravity in space.**   **Your weight depends on gravity.**

Gravity is a really important force to know about. Gravity keeps things on Earth. It also keeps Earth circling the sun. It keeps the moon circling Earth. And it keeps our bodies on this planet! Without gravity, we would float around like the food and astronauts in the space capsule.

Now You Know

Main Idea 3: Gravity
Gravity is the force that keeps us on Earth.

Build Your Vocabulary

Vocabulary Review

Use the word bank to complete each statement.

1. A pulling force between two objects is called _____.

2. A change in position is called _____.

3. _____ is a force that occurs when one object rubs against another.

4. The location of an object is its _____.

5. A _____ is any push or pull.

force

friction

gravity

motion

position

Word Play: Idioms

Idioms are phrases that have special meanings.

food for thought

This means "an idea to think about," not "food to eat."

1. Explain what you think *what goes up must come down* means.

2. Which vocabulary word best fits *what goes up must come down*? Why?

3. Explain what you think *you rub me the wrong way* means.

4. Which vocabulary word best fits *you rub me the wrong way*? Why?

Check Your Understanding

Show What You Know

Main Ideas: **Write the answer to each question.**

1. What does it take to make an object move?

2. What does friction do to an object that is moving?

3. What is the force that keeps us on Earth?

Critical Thinking

1. **Analyze** Why do you think speed skaters wear such tight-fitting skating suits?

2. **Synthesize** Why do you think astronauts are often attached with a line to the spacecraft when they take space walks?

Writing in Science

Write a paragraph Describe a time when friction was useful to you.

- ◆ Use at least two science vocabulary words.
- ◆ Put details in your paragraph.
- ◆ Remember, a good paragraph includes
  - a topic sentence,
  - a body, and
  - a closing sentence.

Process Skill — Quick Activity

Identify variables There are many experiments being done on the *International Space Station*. Some involve growing plants. Can you think of at least one variable, or thing, that would be different about growing plants in space rather than on Earth?

Research the variable you chose. Explain how it might affect plant growth.

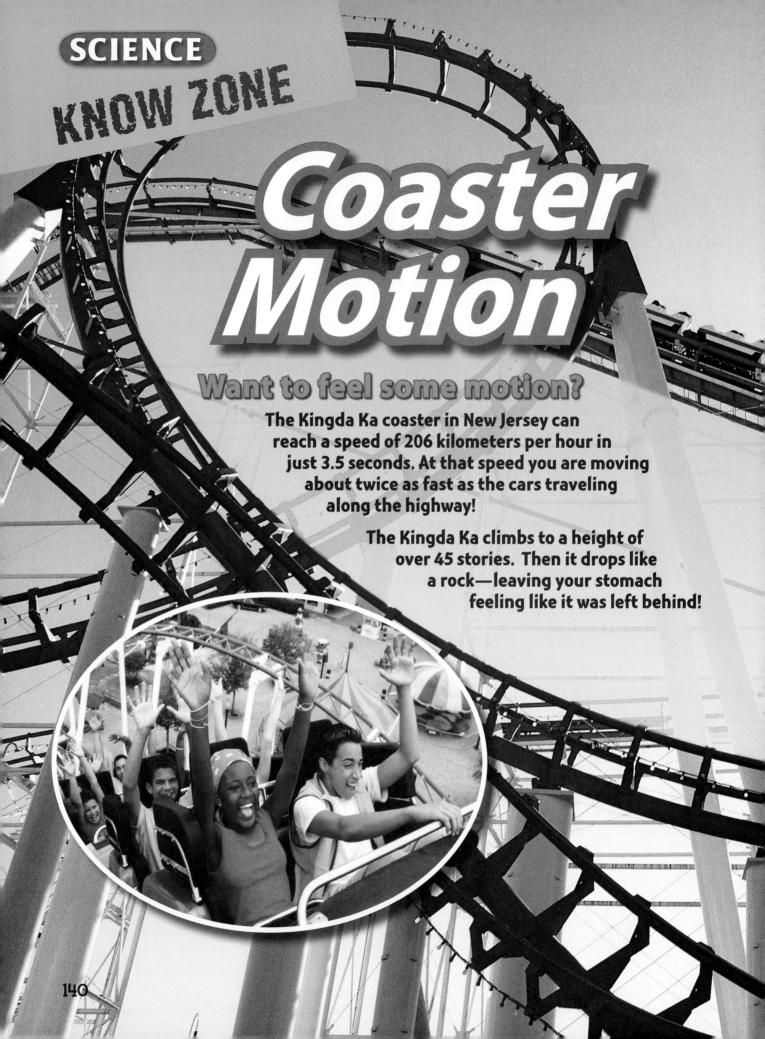

Coaster Motion

Want to feel some motion?

The Kingda Ka coaster in New Jersey can reach a speed of 206 kilometers per hour in just 3.5 seconds. At that speed you are moving about twice as fast as the cars traveling along the highway!

The Kingda Ka climbs to a height of over 45 stories. Then it drops like a rock—leaving your stomach feeling like it was left behind!

CHALLENGE ZONE

Critical Thinking

● Do you think roller coaster designers are interested in friction? Why?

● How do you think gravity helps to increase the thrill on a roller coaster?

Find Out More!

Research on the Web
Find out about the safety of roller coasters. What do designers do to keep people safe? Why do you have to be a certain height to ride some roller coasters?

A Lot of G's!

- A unit that describes the force that we feel is called the g. Earth's gravity is equal to 1 g.

- Astronauts blasting off in the space shuttle feel about 3 g's.

- People riding the Titan roller coaster at Six Flags in Texas feel about 4.5 g's of force!

Magnetism

▶ Play Video: Introduction

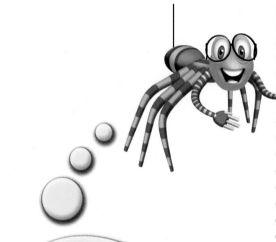

What can magnets do?

New Vocabulary

attract to pull toward itself

electric current electricity that flows through a circuit

electromagnet a temporary magnet created when current flows through wire wrapped around an iron bar

magnetism a force that exists around a magnet and exerts a pull on most metals

repel to push away

Soon You'll Know

Main Ideas

1. How magnets act
2. Some characteristics of magnets
3. How electricity and magnetism are related

Remember In the video you learned about magnets. Now you know what to use the next time you have to pick up a bunch of spilled paper clips!

Think about it Have you ever had loud hiccups in class? If so, those hiccups might have caused you to **attract** some attention. Magnets attract things too. But not by getting the hiccups! Magnets attract physical objects to themselves.

Magnets come in many shapes and sizes.

Many things are not attracted to magnets.

Most metals are attracted to magnets.

A force called **magnetism** exists around magnets. Magnetism exerts a pull, but only on certain metals such as iron and steel. That's why a magnet can pick up a metal paper clip. But it can't pick up a rubber ball. A magnet won't pick up a penny either.

Now You Know

Main Idea 1: Magnets

Magnets attract most metal objects.

Remember In the video you saw how opposites attract. But we're not talking about a magnet's personality. We're talking about its **magnetic field.**

Think about it The **poles,** or ends, of a magnet are where the magnetic force is the strongest. But the poles on magnets don't attract all other magnetic poles. The north poles of two magnets will **repel,** or push away from, each other.

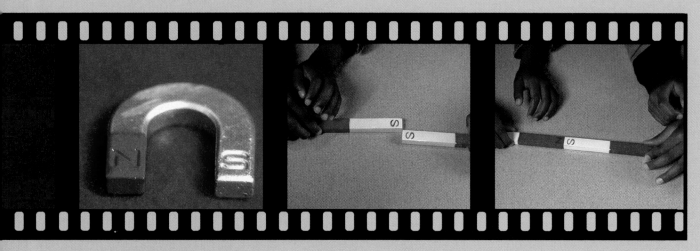

| A magnet's ends are its poles. | Like poles repel. | Opposite poles attract. |

But if you put the north pole of one magnet near the south pole of the other, there is immediate attraction! They will snap together. As the saying goes, opposites attract. Opposite poles attract each other. Like poles repel each other.

Now You Know

Main Idea 2: Magnetic Poles

A magnet's ends are called poles. Like poles repel and unlike poles attract.

▶ **Play Video C: Electromagnets**

Remember In the video you learned about electromagnets. Were you surprised how many times a day you use one?

Think about it You plug a refrigerator into the wall. **Electric current** flows through wires. Your food stays cold. But that's not all an electric current can do in a refrigerator. It can also make a temporary magnet that keeps the refrigerator door shut tightly.

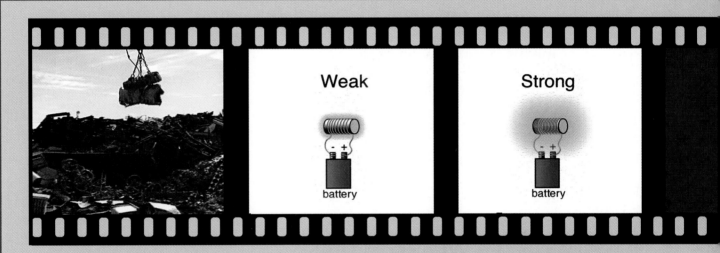

Current flows through electromagnets.

The more wires looped around the iron bar, the stronger the electromagnet.

When an electric current flows through wire that is wrapped around an iron bar, an **electromagnet** is made. An electromagnet attracts metal when the current is flowing through it. When the current stops, the magnetism stops too. Electromagnets are in stereos, telephones, and cars.

Now You Know

Main Idea 3: Electromagnets

Electric current flowing through a wire looped around an iron bar can make an electromagnet.

Build Your Vocabulary

Vocabulary Review

Use the word bank to complete each statement.

1. _____ is the force that exists around a magnet and exerts a pull on most metals.

2. A magnet made when current flows through wire wrapped around an iron bar is an _____.

3. To _____ means to push away.

4. To _____ means to pull toward itself.

5. Electricity that flows through a circuit is an _____.

attract

electric current

electromagnet

magnetism

repel

Word Study: Word Roots

Many parts of science words come from other languages. Those parts are called roots.

The root *electri* or *electro* means "an alloy of gold and silver" in Latin. The vocabulary word *electromagnet* uses this root.

| electro | + | magnet | = | *electromagnet* |

1. Look in your dictionary for three more words that use the root *electri*.

2. Write each word and what it means.

3. Write a sentence using each word.

Check Your Understanding

Show What You Know

Main Ideas: Write the answer to each question.

1. What do magnets attract?

2. What are the ends of a magnet called, and what do they repel?

3. How can an electromagnet be made?

Critical Thinking

1. **Evaluate** Explain why you think it might be useful to use an electromagnet to make a doorbell work.

2. **Apply** Explain why a magnet might be a helpful tool to have in your home.

 Math **in Science**

Calculate A box of paper clips spilled in class. Sue used a magnet to pick up 12 paper clips. Jim's magnet picked up 8. Keisha's magnet picked up 10 paper clips. There are still 5 paper clips on the floor.

♦ How many paper clips did all 3 magnets pick up?

♦ How many paper clips spilled on the floor?

Process Skill **Quick Activity**

Infer Ben put the ends of two magnets together. The magnets did not attract each other. Do you think Ben put like or opposite poles together? What could he do to make the magnets stick together?

Alia wanted to hang a photo on the cabinet door. She used a magnet, but the photo and the magnet fell to the floor. What can you infer?

Simple Machines

▶ **Play Video: Introduction**

What are simple machines?

New Vocabulary

inclined plane a flat surface that is raised at one end

lever a straight bar that moves on a fixed point

pulley a simple machine that uses a wheel and rope to lift a weight

screw an inclined plane wrapped into a spiral

wedge two inclined planes placed back-to-back

wheel and axle a wheel that turns on a post

Soon You'll Know

Main Ideas

1. **Why we use machines**
2. **How levers, inclined planes, and screws are used**
3. **How wedges, pulleys, and wheels and axles are used**

▶ **Play Video A: Simple Machines**

Remember In the video you learned about machines. Isn't it nice to know that so many machines are simple? Sometimes simple things are the most useful.

Think about it You would never pound nails into wood without a hammer. Why? Because you know there is a much better way to do things. A hammer is a simple machine that will make your job much easier.

We use simple machines every day.

Simple machines help us work and play.

One nice thing about a simple machine is that you never have to worry about where to plug it in! Simple machines don't use electricity. Simple machines have few or no moving parts. But they still can help get work done. Simple machines can even help you play.

Now You Know

Main Idea 1: Simple Machines

Machines are tools that make work easier to do. Machines with few or no moving parts are simple machines.

Remember In the video you learned about three simple machines: levers, inclined planes, and screws.

Think about it Squirrels might use their teeth to crack nuts. But people usually have better luck with a simple machine called a nutcracker. That way they don't break their teeth! A nutcracker is made of two **levers** joined together. The levers break the hard shell of the nut apart so we can get to the good stuff inside.

| A ramp is an inclined plane. | A tire jack is a lever. | A screw can hold things together. |

An **inclined plane** can make loading and unloading a truck a lot easier. You might have heard of an inclined plane called a ramp. A **screw** is a special kind of inclined plane wrapped into a spiral. A screw can hold pieces of wood together without a single squirt of glue!

Now You Know

Main Idea 2: Levers and More

Levers and inclined planes make things easier to move. Screws hold things together.

▶ **Play Video C: Wedges and More**

Remember In the video you learned about a wedge, a pulley, and a wheel and axle. These are three more simple machines that help make life much easier!

Think about it What do you get when you put two inclined planes together? You get a simple machine called a **wedge.** What does a wedge look like? One example is the front end of a boat. The boat cuts through the water.

A pulley makes things easier to lift.

The front of this boat is shaped like a wedge.

A steering wheel is a wheel and axle.

Another simple machine is a **pulley.** If you have a tall flagpole at your school, someone probably raises and lowers the flag with a pulley. And let's not forget about the **wheel and axle.** You couldn't steer a car without this simple machine in place!

Now You Know

Main Idea 3: Wedges and More

Wedges, pulleys, and wheels and axles are simple machines that make difficult jobs much easier to do.

Build Your Vocabulary

Vocabulary Review

Use the word bank to complete each statement.

1. A straight bar that moves on a fixed point is a _____.

2. An _____ is a flat surface that is raised at one end.

3. A _____ is a simple machine that uses a wheel and a rope to lift something.

4. Two inclined planes placed together make a _____.

5. A _____ is a wheel that turns on a post.

6. A _____ is an inclined plane wrapped into a spiral.

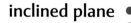

inclined plane

lever

pulley

screw

wedge

wheel and axle

Word Study: In Your Own Words

1. Write the words below on one side of an index card.

| pulley | wedge | wheel and axle |

2. On the back of each card, write a definition in your own words. Draw a picture of each simple machine.

3. Write a sentence using each word.

Check Your Understanding

Show What You Know

Main Ideas: Write the answer to each question.

1. What are simple machines?

2. How would you use a lever, an inclined plane, and a screw to do work?

3. What is an example of a wedge, a pulley, and a wheel and axle?

Critical Thinking

1. **Analyze** What is a piece of playground equipment that uses a simple machine?

2. **Synthesize** How do you think ancient people were able to build huge stone pyramids?

Writing in Science

Quick write Explain the difference between a lever and a pulley.

♦ Draw a diagram of each simple machine.

♦ Tell which of these simple machines you would use to open a can of paint.

Process Skill — Quick Activity

Make a model A teeter-totter is a type of lever. You can balance both sides of a teeter-totter by having equal weight on each end. Using items around your classroom, make a model of a teeter-totter. Can you get the two sides to balance? What can you do if you have different weights on each end?

Multiple Choice Practice Read each question. Choose the best answer.

1 The north-seeking pole of a magnet is labeled ____.

A N
B S
C E
D W

3 The magnets in the diagram are ____.

A not magnetic
B electromagnets
C repelling
D attracting

2 The location of an object at a particular point in time is its ____.

A motion
B magnetism
C position
D gravity

4 ____ is a flat surface that is raised at one end.

A a wheel and axle
B an inclined plane
C a screw
D a wedge

Extended Response Practice Write your answer on a separate piece of paper.

5 How does friction affect movement?

STOP

CHAPTER 8

Matter

Getting Started | Quick Activity

- **Discussion** Can you believe the butterfly in the picture is made totally of ice? Look at the things around you. What kind of materials make up these things? Can you think of an invisible thing in the room right now?

- **Critical Thinking** What do you think will happen to the butterfly when the weather gets warmer?

Properties of Matter

▶ **Play Video: Introduction**

What is matter?

New Vocabulary

mass the measure of the <u>amount</u> <u>of matter</u> in an object

matter anything that takes up <u>space</u> and has <u>mass</u>

property something you can observe <u>with your senses</u>

texture how the surface of an <u>object</u> feels to the touch

volume the measure of <u>how much</u> <u>space</u> matter takes up

Soon You'll Know

Main Ideas

1. **What matter is**
2. **How to describe matter**
3. **How some properties of matter are measured**

▶ **Play Video A: Matter**

Remember The video you just watched was about matter. Of course, you already know something about the subject. Why? Because you are made of matter too!

Think about it Have you ever accidentally bumped into a door? If so, you know that a door is made of **matter.** This means it takes up space and has mass. **Mass** is the amount of matter in an object. A balloon doesn't have a lot of mass. A door has a lot of mass.

| These coins are made of matter. | The balloon is made of matter. So is the air inside it. | Weight can change, but mass stays the same. |

Mass and weight are not the same. Weight depends on the pull of gravity. So an object's weight can change depending on where that object is. But the mass of an object always stays the same. Your weight on Earth would be different from your weight on the moon. Your mass would be the same in both places.

Now You Know

Main Idea 1: Matter

Matter is anything that takes up space and has mass.

▶ Play Video B: Properties of Matter

Remember In the video you learned about the properties of matter. From gooey to gruesome, the properties of matter make the world an interesting place!

Think about it Pretend you just lined up your shoes and your friends' shoes along the wall. How would you describe them? Big? Small? Colorful? Smelly? You can probably think of lots of different ways to talk about them. That's because all matter, including shoes, can be described by **properties.**

Sweet, fresh, round, and red are all properties.

The ice cube is cold.

The snail is slimy.

We observe the properties of matter with our senses. We might say something is hot or cold. We might say the fruit tastes sweet. We might even describe the **texture,** or feel, of an object. Different matter has different properties so it's described in different ways. Shape, size, color, and texture are all properties. And there are many more.

Now You Know

Main Idea 2: Properties of Matter

Matter can be described using properties.

Remember In the video you learned that mass and volume can be measured. Good thing too. You wouldn't want to scuba dive without knowing how much air you have in your scuba tank!

Think about it All matter can be measured. That's because matter has mass and takes up space. So, for example, if you want to find the mass of a book, you can place it on a metric balance. The balance will measure the book's mass in grams.

You can measure length in centimeters.

Volume is a measure of the amount of space something takes up.

But you can also measure matter by its **volume,** or the amount of space it takes up. It's easier to measure liquid this way. If you pour water into a measuring cup, the mark at the top of the water level will show you its volume.

Now You Know

Main Idea 3: Mass and Volume

Mass and volume are properties of matter that can be measured.

Build Your Vocabulary

Vocabulary Review

Use the word bank to complete each sentence.

1. A _____ is a characteristic of matter and can be observed with your senses.

2. A word used to describe how the surface of an object feels to the touch is _____.

3. The measure of the amount of matter in an object is its _____.

4. The measure of how much space matter takes up is its _____.

5. _____ includes anything that takes up space and has mass.

mass

matter

property

texture

volume

Word Play: Rhyming

Poems and songs often use words that rhyme, or have the same sounds at the end. For example:

> Rusty, smooth, rough, and impossible to shatter.
>
> These words describe the properties of matter!

Now you do it. Write your own poem using one or more of the new vocabulary words. Make the last two words rhyme.

Check Your Understanding

Show What You Know

Main Ideas: Write the answer to each question.

1. What is a good way to define matter?

2. What are some of the properties of matter?

3. What are two properties of matter that can be measured?

Critical Thinking

1. **Apply** How could you use properties of matter to help you organize your toys?

2. **Analyze** Explain how we know air is matter even though we can't see it.

Writing in Science

Write a description Describe your favorite meal. Think about the different properties of the food. Try to use all of your senses as you write.

♦ How does the food smell?

♦ What does it look like?

♦ How does it taste?

♦ You can even write about how your favorite food makes you feel!

Process Skill — Quick Activity

Classify When you classify, you group things together by their properties. List ten things in your classroom. Decide how you would classify these objects into two big groups.

Now share your list with a partner. Ask your partner to divide the list into two groups. Did you and your partner use the same properties to classify these objects?

Solids, Liquids, and Gases

▶ **Play Video: Introduction**

How can matter change?

Soon You'll Know

Main Ideas

1. About three states of matter
2. What happens when matter undergoes a physical change
3. What happens when matter undergoes a chemical change

▶ **Play Video A: States of Matter**

Remember In the video you learned that matter comes in different states. The fizzy stuff in soda is in one state. The soda itself is in a different state. And the cup holding it all is a third state of matter.

Think about it Pink lemonade, a rock, and oxygen gas are all forms of matter. But why do they seem so different? They are examples of matter in different states.

Matter can be solid, liquid, or gas.

Liquid takes the shape of its container.

Gas spreads out in the air or fills a balloon.

The bottom line is that most matter exists in one of three states. A **solid** has a definite shape and volume. A **liquid** has a definite volume but no definite shape. And a gas—well, that's the loosest one of them all. A **gas** has no definite volume and no definite shape.

Now You Know

Main Idea 1: States of Matter

Matter exists in three states: solids, liquids, and gases.

163

▶ **Play Video B: Physical Changes**

Remember In the video you learned about physical changes of matter. Let's face it. No matter how many ways you slice it, a pizza is still a pizza!

Think about it Have you ever played with modeling clay? You can stretch it and roll it. You can flatten it and stack it. But no matter how you change its shape, the clay is still clay. A pizza is still a pizza even after you cut it.

Sugar in a cube or crushed is still sugar.

Cutting paper and chopping tomatoes are physical changes.

When you shape clay, you are making a **physical change** to it. When you cut paper, you make a physical change. When you freeze water, you make a physical change. In a physical change, matter changes only in size, shape, or form. The type of matter itself does not change.

Now You Know

Main Idea 2: Physical Changes

In a physical change, matter only changes in size, shape, or form.

▶ **Play Video C: Chemical Changes**

Remember In the video you learned about chemical changes. These kinds of changes produce some pretty tasty treats in the kitchen!

Think about it It seems like magic. A cake goes into the oven as liquid batter. A little while later, it comes out of the oven as a solid cake. Through heat, one kind of matter has been changed into another kind of matter. But this is no magic trick. It's just a **chemical change.**

Rusting is a chemical change.

A chemical change occurs when a cake is baked.

You know that in a physical change matter remains the same. But things are different in a chemical change. In this kind of change, one or more substances actually change to different substances. This means the matter has different properties than it did before.

Now You Know

Main Idea 3: Chemical Changes

In a chemical change, one kind of matter becomes a different kind of matter.

165

Build Your Vocabulary

Vocabulary Review

Use the word bank to complete each statement.

1. A _____ has a definite volume but no definite shape.

2. A _____ has no definite shape and no definite volume.

3. A _____ has a definite volume and a definite shape.

4. A change that forms a different kind of matter is a _____ change.

5. A change in the way matter looks that leaves the matter itself unchanged is a _____ change.

chemical

gas

liquid

physical

solid

Word Study: Classifying

Draw these circles and labels on your paper.

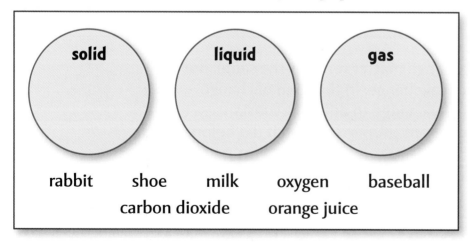

solid   liquid   gas

rabbit   shoe   milk   oxygen   baseball
carbon dioxide   orange juice

1. Read each term listed above.

2. Decide if it is a solid, liquid, or gas.

3. Write the term in the correct circle on your paper.

Check Your Understanding

Critical Thinking

1. **Analyze** Is it a chemical or physical change when an ice cube melts? Why?

2. **Synthesize** Explain what kind of changes happen to a banana when you eat it.

 Words **in Science**

Process Skill **Quick Activity**

Academic Vocabulary Use a dictionary to look up the word *state*.

♦ What are the guide words on the page? How many meanings does the word *state* have?

♦ Find the word *state* in the vocabulary definitions.

♦ Which meaning best fits the word *state* in the definitions? Why?

Communicate Cut out photos of different items from newspapers and magazines. Look for examples of solids, liquids, and gases. With a classmate, make a poster that explains the different states of matter. Glue on the photos, and add drawings. Now write what makes each state of matter different from the others.

Way Cool!

Astronauts wear space suits for lots of reasons. One reason is to protect their bodies from high temperatures.

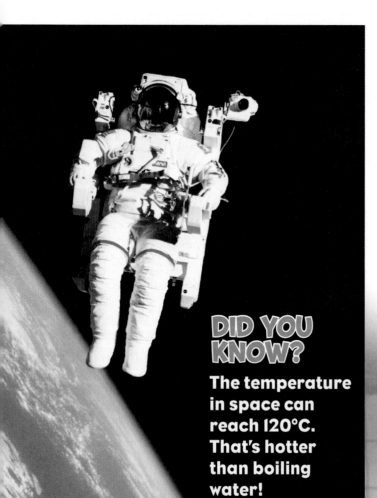

Firefighters wear fireproof suits when they fight a fire. This clothing is heavy and hot. But suits could be different in the future. Help may be coming from a far out place—the astronauts in space!

DID YOU KNOW?

The temperature in space can reach 120°C. That's hotter than boiling water!

Keeping Them Cool!

Spacesuits have cooling tubes built in. These same kinds of tubes could be put into the material of firefighting suits. Water would flow through the tubes to cool the firefighter down.

CHALLENGE ZONE

Critical Thinking

● Why else might an astronaut need to wear a space suit?

● Why do you think it's important to make firefighter suits as lightweight as possible?

Find Out More!

Research on the Web

Race car drivers need protection from fire too. And so do scientists who study volcanoes. Find out what kind of fireproof suits they wear.

Heating and Cooling Matter

▶ **Play Video: Introduction**

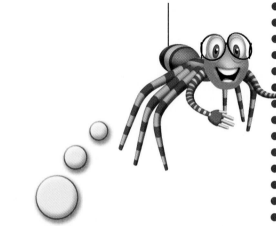

How does heat move?

New Vocabulary

conductor a material that heat travels through easily

freeze to turn from a liquid into a solid

heat the flow of energy from warmer matter to cooler matter

insulator a material that heat doesn't travel through easily

melt to change from a solid to a liquid

Soon You'll Know

Main Ideas

1. **How heat moves through matter**
2. **What properties make good conductors**
3. **What makes a good insulator**

▶ **Play Video A:** Heat Energy

Remember In the video you learned about heat. Now you know it's the flow of heat energy that can really make you shiver or sweat.

Think about it You come inside on a cold winter day. A cup of hot chocolate is waiting for you. At first it feels warm in your hands. Then—ouch! You realize the cup is too hot, and you quickly move your hands. What happened here?

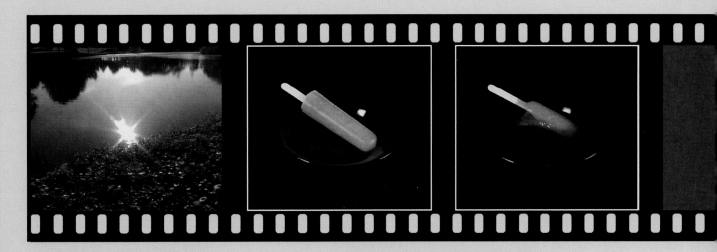

The sun produces enough heat energy to warm Earth.

Heat energy is flowing into the frozen treat, causing it to melt.

You see, **heat** energy moves from a warmer object to a cooler one. It happens all the time. When a frozen treat **melts,** heat energy moves into it. This changes the frozen treat from a solid to a liquid. Do you want to make something **freeze?** Then you have to take enough heat energy away from a liquid to change it into a solid.

Now You Know

Main Idea 1: Heat Energy

Heat is the flow of heat energy from a warmer object to a cooler one.

Remember In the video you learned about good conductors. Conductors transfer **energy,** such as electricity, sound, and even heat.

Think about it If you stir a pot of soup on the stove with a metal spoon—watch out! It may be more than the soup that gets warm. The metal spoon will heat up too, so be careful. But why does this happen?

Metal is a good conductor.   Plastic and wood are not good heat conductors.

The reason the spoon heats up is because it is made of metal. And metal is a good **conductor.** This means heat energy can easily travel through it. But other materials, such as wood, are not good conductors. Do you want to keep your hand safe from the heat? Stir that soup with a wooden spoon!

Now You Know

Main Idea 2: Conductors

Good conductors are materials through which heat moves easily.

Remember In the video you learned about insulators. They're handy for keeping heat in or out. You don't want heat getting into your refrigerator now, do you?

Think about it A bird sits on a snowy branch in the dead of winter. The temperature is freezing. The bird has no coat or hat on, but it seems to be quite comfortable. How does this tiny creature keep itself so warm? Its feathers are good **insulators.**

A layer of fat is a good insulator.

Insulation keeps our houses warm.

Oven mitts insulate your hands from heat.

You already know that heat moves easily through a good conductor. But an insulator does the opposite. Good insulators don't let heat move through them very easily. For example, a thick oven mitt is a good insulator. It won't let heat from a hot pan reach your hand. Likewise, a good winter coat will keep heat near your body.

Now You Know

Main Idea 3: Insulators

Good insulators are materials that do not let heat move through them easily.

Build Your Vocabulary

Vocabulary Review

Use the word bank to complete each statement.

1. When you _____ something, it changes from a liquid to a solid.

2. The flow of energy from warmer matter to cooler matter is _____.

3. A material through which heat travels easily is a good _____.

4. When you _____ something, it changes from a solid to a liquid.

5. A material through which heat does not travel very easily is called an _____.

conductor

freeze

heat

insulator

melt

Word Study: Word Forms

Some nouns can be made into verbs by changing the letters in the word.

Change the nouns into verbs by adding, removing, or changing letters. The first one is done for you.

Noun	Verb
1. conductor	conduct
2. insulator	
3. heater	
4. freezer	

174

Check Your Understanding

Show What You Know

Main Ideas: Write the answer to each question.

1. What is heat, and how does it move?

2. What types of matter are good conductors?

3. What does a good insulator do?

Critical Thinking

1. **Analyze** Explain why it's important for firefighters to have gloves and boots with good insulation.

2. **Apply** Explain why a wooden swing might be more comfortable in the summer than a metal swing.

 Words **in Science**

Academic Vocabulary Look up the word *material* in a dictionary.

♦ What are the guide words on the page?

♦ How many meanings are listed for *material*?

♦ Which meaning best fits the way *material* is used in this lesson? Why?

Process Skill **Quick Activity**

Experiment Plan an experiment to test what materials will keep an ice cube frozen longest. Write down and gather the materials. Wrap your ice cube as you planned. After two hours, check your ice cube. Record your observations. Rewrap it and check it again in two hours. Repeat until the ice cube is gone. Compare your results with the class. What can you conclude?

 TIPS

Making Comparisons Comparisons tell how two things are alike or different from each other. Some tests will ask you to compare information on charts or diagrams.

Multiple Choice Practice Read each question. Use the chart to choose the best answer.

Substance	Temperature	State of Matter
Water	25°C	Liquid
Water	0°C	Solid
Water	100°C	Gas

1 Water can exist in which state?

A solid
B liquid
C gas
D all of the above

3 Ice is water as a solid. What temperature is ice?

A 25°C
B 0°C
C 100°C
D 110°C

2 Which state of water has the highest temperature?

A gas
B liquid
C solid
D all of the above

4 Water in a liquid state is _____ than water as a gas.

A hotter
B warmer
C cooler
D less dense

Extended Response Practice Write your answer on a separate piece of paper.

5 Compare water vapor and liquid water.

 STOP

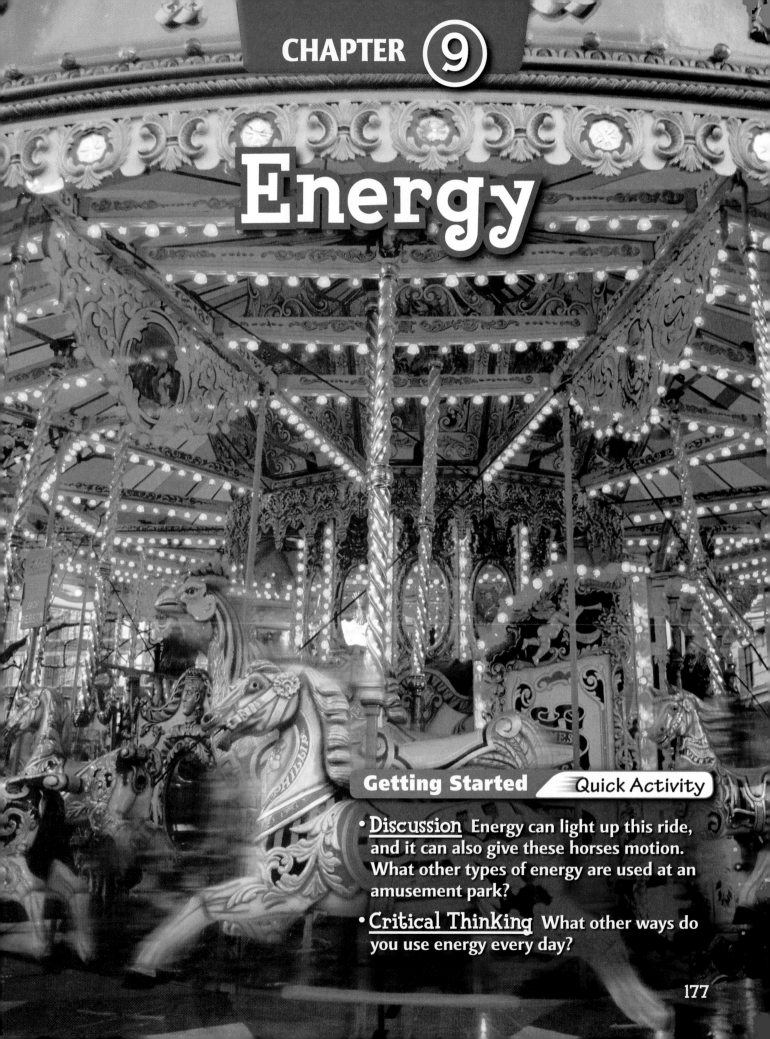

Energy

Getting Started *Quick Activity*

- **Discussion** Energy can light up this ride, and it can also give these horses motion. What other types of energy are used at an amusement park?

- **Critical Thinking** What other ways do you use energy every day?

Light and Sound

▶ **Play Video: Introduction**

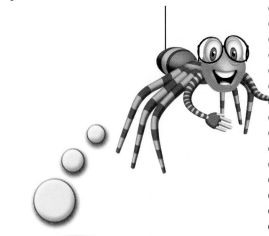

What are light and sound?

New Vocabulary

pitch how <u>high</u> or how <u>low</u> a sound is

prism a <u>block</u> of angled <u>glass</u> that bends the waves of light passing through it

reflect to <u>bounce</u> light off of a surface

vibrate to move <u>back and forth</u> quickly

volume how <u>loud</u> or <u>soft</u> a sound is

Soon You'll Know

Main Ideas

1. **What light is and how it travels**
2. **What happens when light strikes an object**
3. **What sound is and how it is created**

▶ **Play Video A: Light Waves**

Remember In the video you learned how light energy travels. So catch a wave and read on!

Think about it There are many sources of light in this world. Some, like the sun and lightning, are natural sources. You can see them in all kinds of places. Others, like a candle and a lightbulb, are the results of people's bright ideas. But no matter where it comes from, all light is a form of energy.

| We depend on light from the sun. | Light reflects well off a mirror. | Light waves pass through a clear glass of water. |

So what happens when light waves strike an object? It all depends on the material the light waves hit. Some waves may be absorbed by the object. Sometimes the waves go into or through the object, like a window. And sometimes the waves **reflect** off the object, like a shiny penny or a mirror.

Now You Know

Main Idea 1: Light Waves

Light is energy that travels in waves.

▶ **Play Video B: Color**

Remember In the video you learned about light waves and color including red, orange, yellow, green, blue, indigo, and violet.

Think about it It's really an amazing sight. White light strikes a **prism.** The light waves are bent, or *refracted,* and a pattern of rainbow colors appears. Think about when you see a rainbow after a storm. Each raindrop in the air acts like a tiny prism. They all bend the light, and you see a rainbow.

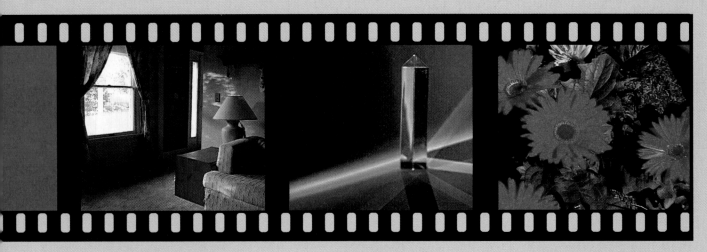

Sunlight passes through a window.

A prism bends light waves.

The red flower reflects red light.

So if light is made up of so many colors, how come everything doesn't look like a rainbow? The color of an object depends on how it reflects or absorbs light. A red flower absorbs all the colors in light except red. The red light waves reflect off the flower, so that's the color we see.

Now You Know

Main Idea 2: Color

Light can pass through an object, be reflected, or be absorbed.

▶ **Play Video C: Sound Waves**

Remember In the video you learned about sound waves. You can't see them, but you can probably hear them loud and clear!

Think about it The **wind** blows a door shut. This causes the door and the air around it to **vibrate.** These vibrations make sound waves. The sound waves travel to your ear. Nerve cells in your ear send the information to your brain—and SLAM! You hear the sound!

Sound waves are captured by your ear.

Instruments vibrate and create sound waves.

Sound waves with a lot of energy are loud.

But not all sounds are the same. Slow vibrations make sounds with a low **pitch.** Fast vibrations make the pitch go higher. Do you want to make a loud sound? Give a shout! That puts a lot of energy into the sound wave. If you want to turn down the **volume,** then whisper. Less energy in the sound waves means less volume.

Now You Know

Main Idea 3: Sound Waves

Sound is made when an object vibrates and creates sound waves.

Build Your Vocabulary

Vocabulary Review

Use the word bank to complete each statement.

1. A block of angled glass that bends waves of light passing through it is a _____.

2. _____ tells you how high or low a sound is.

3. To _____ means to move back and forth very quickly.

4. _____ tells you how loud or soft a sound is.

5. Light can _____, or bounce off, a surface.

pitch

prism

reflect

vibrate

volume

Word Study: In Your Own Words

1. Write the words below on one side of an index card.

pitch	vibrate	prism

2. On the back of each card, write a definition in your own words.

3. Write a sentence using each word.

Check Your Understanding

Show What You Know

Main Ideas: Write the answer to each question.

1. What is light, and how does it travel?

2. What can happen when light waves strike an object?

3. How is sound made, and what affects its volume?

Critical Thinking

1. **Synthesize** Describe what you think is happening to light when you see the color black.

2. **Apply** Explain why the sound of a train is louder than the sound of a car horn.

Writing in Science

Write a description Describe your favorite type of music. Explain why you like it.

◆ Try to use the words *sound waves*, *volume*, and *pitch* in your description.

◆ Put details in your description.

◆ Use your senses. How does the music sound? Can you feel the music?

Process Skill — Quick Activity

Observe Hold a plastic ruler so that one-half of it hangs off the edge of your desk. Tap the end of the ruler. Describe what you hear. Record your observations in a chart.

Now move the ruler so that one-fourth of it is hanging over the edge. Repeat the activity. Do the same thing with three-fourths of the ruler over the edge. What can you conclude?

Lighting the Sky

Nature puts on its own fireworks show when lightning streaks across the sky. But make no mistake, lightning is an electrical charge. Those bright flashes may be beautiful, but they are very dangerous too.

Follow the 30/30 Rule

30 Seconds

1. When you see lightning, count the time until you hear thunder.
2. If you count 30 seconds OR LESS, lightning is near enough to hurt you.
3. Take cover inside a shelter immediately!

30 Minutes

Wait 30 minutes after seeing the last lightning flash before leaving the shelter!

True or False?

Lightning never strikes the same place twice.

BEEP ... that's false! Lightning often strikes in the same place. The Empire State Building in New York City gets struck by lightning about 100 times a year.

DID YOU KNOW?

A lightning flash can be from 8,000°C to 33,000°C. That's hotter than the surface of the sun!

BAD NEWS	GOOD NEWS
Lightning can kill people. In the United States about 80 people are struck and killed by lightning each year.	Not everyone who is struck by lightning dies!
Lightning can start fires. In the United States about 10,000 fires are started by lightning each year.	Lightning puts nitrogen in the ground for plants to use. This is very helpful in the cycle of life

CHALLENGE ZONE

Critical Thinking

- Why do you think it is not safe to stand under a tree during a thunderstorm?

- Why should an outside sporting event be cancelled if lightning starts to flash during the game?

Find Out More!

Research on the Web

Find out more about lightning strikes. Which state gets the most cloud-to-ground lightning?

Electrical Energy

▶ **Play Video: Introduction**

New Vocabulary

battery a source of electric current

circuit a complete path through which electricity can flow

electric current electricity that flows through a circuit

switch a device that can open or close an electric circuit

What is electricity and how do we use it?

Soon You'll Know

Main Ideas

1. What electricity is and how we use it
2. How electric current moves
3. How electricity is controlled

Remember In the video you learned about electricity, or electrical energy. It's shocking how far we've come from the first lightbulb!

Think about it A bolt of lightning is a powerful source of electrical energy. However, the power behind other sources of electricity may not be so clear. A plug in a wall looks very simple. But that plug is connected to a huge power supply.

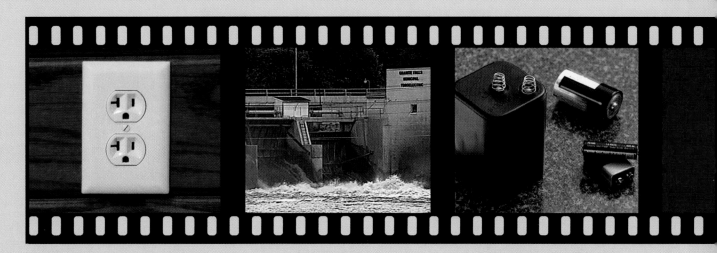

Electricity powers many things in homes.

This power plant uses water to make electricity.

Batteries power many devices.

Nuclear energy, coal, and water are all used to make electricity in power plants. Those plants send electricity into your home and school. But what if you need power on the go? That's when a **battery** comes in handy. Batteries change chemical energy into electrical energy. They power things such as cell phones and MP3 players.

Now You Know

Main Idea 1: Electricity
Electricity is a form of energy.

Play Video B: Circuits

Remember In the video you learned how electricity moves. It's all about creating the right path!

Think about it Have you ever followed a path through a zoo or park? The path shows you where to go. It takes you through the zoo and back to where you started. Electricity follows a path too. It's called a **circuit.** Electricity flows through the wires that make up the circuit.

Electricity travels through wires into our homes.

Electricity runs this washing machine and this television.

A closed circuit creates a complete path for **electric current** to follow. If you plug a lamp into a socket and turn it on, you create a closed circuit. This allows an electric current to flow from the power source in the socket through the lamp to the lightbulb. If you unplug the lamp, the circuit is broken. Electric current can no longer flow to the lamp.

Now You Know

Main Idea 2: Circuits

Electric current flows through a circuit.

▶ **Play Video C: Electric Switches**

Remember In the video you learned about switches. They turn on a whole new way of thinking about energy! Turn it on, and turn it off.

Think about it What if the only way to turn on a lamp was to plug it into a socket or outlet? Plugging it in and unplugging it all the time would be a lot of work. Luckily, we don't have to do that. Why? Because we have **switches.**

Lights can be turned on and off.

Switches come in many shapes and sizes, but they all open and close electric circuits.

A switch can open or close an electric circuit. When the switch is on, the circuit is complete. We say the circuit is closed. This means electric current can flow. When the switch is off, the circuit is broken. The electric current will stop.

Now You Know

Main Idea 3: Electric Switches

An electric circuit can be opened or closed with a switch.

Build Your Vocabulary

Vocabulary Review

Use the word bank to complete each statement.

1. A device that can open or close an electric circuit is a _____.

2. _____ is electricity that flows through a circuit.

3. For power on the go, a _____ supplies a source of electric current.

4. A complete path through which electricity can flow is called a _____.

> battery
>
> circuit
>
> electric current
>
> switch

Word Play: Science Riddles

Solve the riddles below using words from the lesson.

Science Riddles

1. I am small and can change chemical energy into electrical energy.

2. I am a lever that can open and close.

3. I am a pathway, and electricity follows me.

4. I light up when electricity flows.

5. I make electricity using resources, such as coal and water.

Check Your Understanding

Main Ideas: Write the answer to each question.

1. What is electricity, and how do we use it?

2. What is a circuit?

3. How does a switch control the flow of electricity?

Critical Thinking

1. **Apply** Explain why all switches don't look the same.

2. **Analyze** Why do you think batteries eventually stop working?

Math in Science

Calculate Jorge needs to put an equal number of batteries in 3 hurricane supply boxes. It takes 2 batteries to light one flashlight.

♦ If he has 18 batteries, how many batteries should Jorge put in each box?

♦ How many flashlights can he use?

Process Skill / Quick Activity

Make a model Suppose you had a lightbulb, a wire, and a battery. Can you think of a way to put them together to get the lightbulb to light?

Draw a diagram of how you would connect your items. Label each part. Now see if you can gather the materials and test your design.

Energy Sources

▶ **Play Video: Introduction**

How do we use different forms of energy?

New Vocabulary

conserve to save, protect, or use something wisely without wasting it

energy the ability to do work

fuel a substance that is burned for its energy

nonrenewable cannot be reused or replaced easily

renewable can be replaced in a short period of time

solar energy energy from the sun

Soon You'll Know

Main Ideas

1. Where we can find stored energy
2. About ways that energy changes
3. About renewable and nonrenewable sources of energy

▶ **Play Video A: Stored Energy**

Remember In the video you learned about stored energy. Stored energy comes in many forms. What's your favorite fuel to eat?

Think about it There's a reason to eat a good breakfast every morning. You work hard in school. And your body needs **fuel** to give you the **energy** to make it to lunch! You need energy to run and play. You also need energy to turn the pages of this book.

Everyone needs energy to live.

Food is a source of fuel for people.

Gasoline is a source of fuel for cars.

Food is a source of stored energy for people. Plants are a source of stored energy for many animals. But there are other sources of stored energy. When cars burn gasoline, they are using a form of stored energy. When people use cell phones, the batteries are the stored energy source.

Now You Know

Main Idea 1: Stored Energy

Sources of stored energy take many forms, such as food, fuel, and batteries.

Remember In the video you learned how energy can change forms. Because we need energy to live, it's important to have a constant supply!

Think about it The stored energy in food doesn't do us much good unless we can change it into the kind of energy that we need to live. Our bodies can change energy from one form to another. For example, some of the stored energy in food can be used to make heat energy to keep us warm.

Stored energy in food can be changed.

Gasoline has stored energy to make cars run.

Batteries can light the night.

Stored energy is useful in other ways too. The energy stored in gasoline can be changed into mechanical energy to make cars run. The stored energy in batteries can be changed to make a lightbulb light or make music play.

Now You Know

Main Idea 2: Energy Changes Forms

Energy can change from one form into another.

▶ **Play Video C: Renewable Energy**

Remember The video you just watched was about renewable energy resources. It's an exciting field of study for the future!

Think about it No doubt you've been hearing for a long time that you should **conserve** energy. That's really important for **nonrenewable** resources because someday they could run out. It takes a very long time for resources like gasoline, oil, and coal to form. And when they are gone, we can't just go out and get more.

We need a constant source of energy.

Wind and solar energy are renewable energy resources.

Luckily, other sources of energy are **renewable,** or replaceable, over a short period of time. Water, wind, and **solar energy** are renewable resources. If we want to have a constant supply of energy, it just makes sense to use more renewable resources whenever we can.

Now You Know

Main Idea 3: Renewable Energy

Renewable energy resources are important for the future quality of life.

195

Build Your Vocabulary

Vocabulary Review

Use the word bank to complete each statement.

1. A _____ resource can be replaced in a short period of time.

2. _____ is a substance burned for its energy.

3. _____ is the ability to do work.

4. To _____ means to save, protect, or use something wisely.

5. Energy from the sun is called _____.

6. A _____ resource cannot easily be replaced.

conserve

energy

fuel

nonrenewable

renewable

solar energy

Word Play: Rhyming

Poems and songs often use words that rhyme or have the same sounds at the end. For example:

Oil and coal give us energy. We use them up each hour.

But wind and solar won't give out as a future source of power.

Now you do it. Write your own poem using one or more of the new vocabulary words. Make the last two words rhyme.

Check Your Understanding

Main Ideas: Write the answer to each question.

1. What are three sources of stored energy?

2. Why is it important that energy can change form?

3. Why are renewable energy resources important?

Critical Thinking

1. **Evaluate** What are some advantages and disadvantages of renewable energy resources?

2. **Synthesize** What do you think would happen if your body could not store energy from food?

Writing in Science

Write a letter to the editor Try to convince people that conserving energy is important.

♦ Use a greeting and a closing.

♦ Clearly state your opinion.

♦ Give reasons that support your opinion.

♦ Tell people what could happen if they don't follow your advice.

Process Skill — Quick Activity

Classify Think about all the ways that you use energy. Make a list. Include things such as eating and riding in a car or on a bus.

After each activity, write down whether the energy source for that activity is renewable or nonrenewable. Which sources of energy do you use the most?

Standardized Test Practice 9

 Using Key Words Key words can help you find the correct answer by giving you important details. Key words include *first*, *last*, *before*, *smallest*, and *better*.

Multiple Choice Practice Read each question. Choose the best answer.

1 Which of these circuits would have a lit lightbulb?

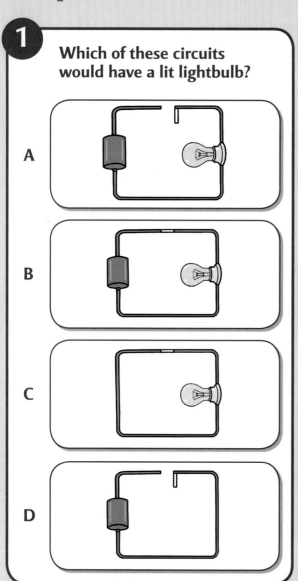

A

B

C

D

2 Which gives off the most light energy?

- A the moon
- B a lightbulb
- C a campfire
- D the sun

3 What happens last in this process?

- A Sound waves hit our ears.
- B Sound waves travel through the air.
- C We hear sound.
- D An object vibrates.

4 Which is not a form of energy?

- A light
- B color
- C heat
- D sound

Extended Response Practice Write your answer on a separate piece of paper.

5 What are two ways you can help conserve energy?

198

SRA Snapshots Video Science™
Science Handbook

Table of Contents

The Metric System

The Metric System

Scientists use the metric system, or the International System of Measurement. The metric system uses different base units for each type of measurement. The base units and the type of measurement they are used for can be seen in the following table:

Type of Measurement	Metric Unit and Symbol	Tool Used to Measure
length	meter (m)	meter stick
volume	Liter (L)	graduated cylinder
mass	gram (g)	balance
time	second (s)	stopwatch or other timer
temperature	degrees Celsius (°C)	thermometer

Even if you don't use these units all the time, they are very easy to understand.

mass = 1 gram

volume = 2 liters

The metric system uses prefixes to create larger or smaller units. Here are a few.

height = 1 meter

Prefix	Symbol	Meaning (numerical)	Meaning (words)
kilo-	k	1,000	thousand
centi-	c	0.01	hundredth
milli-	m	0.001	thousandth

The Metric System

Metric System Conversions

Sometimes you many not have the right tools to perform measurements using the metric system. It is helpful to be able to switch between the standard system and the metric system.

Standard System of Measurement	Multiply by	Metric System of Measurement
if you know **inches**	multiply by 2.54	to get **centimeters**
if you know **feet**	multiply by 0.305	to get **meters**
if you know **miles**	multiply by 1.61	to get **kilometers**
if you know **gallons**	multiply by 3.785	to get **liters**
if you know **pounds**	multiply by .454	to get **kilograms**

Temperature

Converting between the two common temperature units is easy to do. If you know degrees Celsius but want degrees Fahrenheit, use the formula below. Put the temperature you know in place of the °C.

$$°F = (1.8 × °C) + 32$$

If you know degrees Fahrenheit but want degrees Celsius, use the formula below. Put the temperature you know in place of the °F.

$$°C = (°F - 32) ÷ 1.8$$

In **Celsius**, water freezes at 0°C and boils at 100°C. In **Fahrenheit**, water freezes at 32°F and boils at 212°F.

Classification

Kingdoms of Life

All living organisms are grouped into one of six different kingdoms.

Each kingdom of organisms is organized into smaller groups of creatures that are more closely related to one another. As groups contain fewer and fewer organisms, the organisms in those groups are more closely related and are similar to each other.

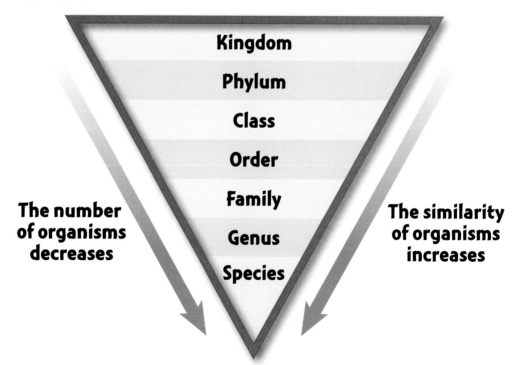

Energy Transfer

Food Web

All organisms depend on other organisms to help provide the things they need. A food web shows how organisms depend on other organisms for food. The arrows point from the organisms being consumed to the organisms that are consuming them.

Energy Pyramid

The organisms in the food web above can be categorized into groups based on how they get the energy they need. Energy flows from the organisms at the bottom of the pyramid to those at the top as lower organisms are consumed by higher organisms.

Amount of food energy decreases

Number of organisms decreases

carnivores
(tertiary consumers)

carnivores and omnivores
(secondary consumers)

herbivores
(primary consumers)

producers

Energy Pyramid

The Planet Earth

Earth's Layers

The planet Earth is divided into several layers. Each layer has different properties.

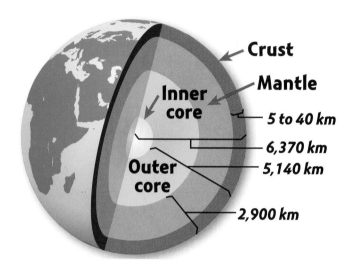

Crust

Mantle

Inner core

Outer core

5 to 40 km

6,370 km

5,140 km

2,900 km

The Water Cycle

The water on Earth's surface is constantly moving through a process called the water cycle.

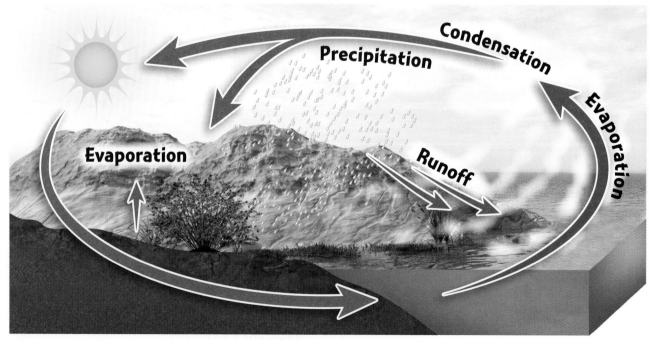

Condensation

Precipitation

Evaporation

Evaporation

Runoff

Earth in Space

Climate Zones

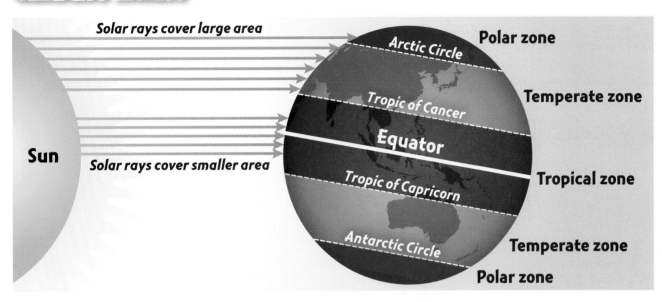

A region's climate, which is the average temperature and precipitation over a long period of time, is greatly affected by its location on Earth. Light and energy from the sun hit Earth at different angles, which results in a different climate. Earth has three primary climate zones: polar, temperate, and tropical.

Eclipses

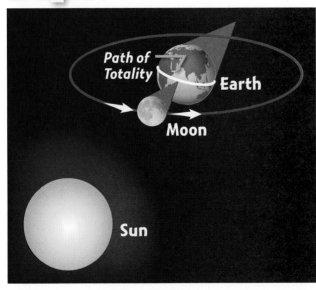

During a **solar eclipse** the new moon's shadow falls on Earth.

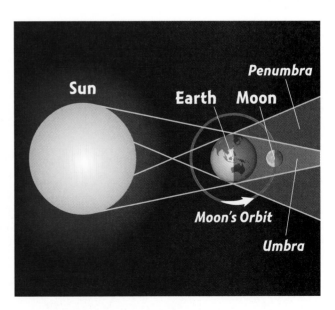

During a **lunar eclipse** the Earth's shadow falls on the full moon.

Electromagnetic Energy

The electromagnetic spectrum shows the different forms of electromagnetic energy.

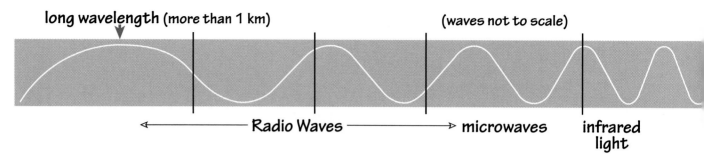

long wavelength (more than 1 km)

(waves not to scale)

← Radio Waves → microwaves

infrared light

Periodic Table of the Elements

The periodic table shows all of the elements, their chemical symbols, and their atomic numbers.

Key

6 ◄· · · · · Atomic number
C ◄· · · · · Element symbol
Carbon ◄· · · · · Element name

State of Matter at 20°C

C Solid
Br Liquid
H Gas

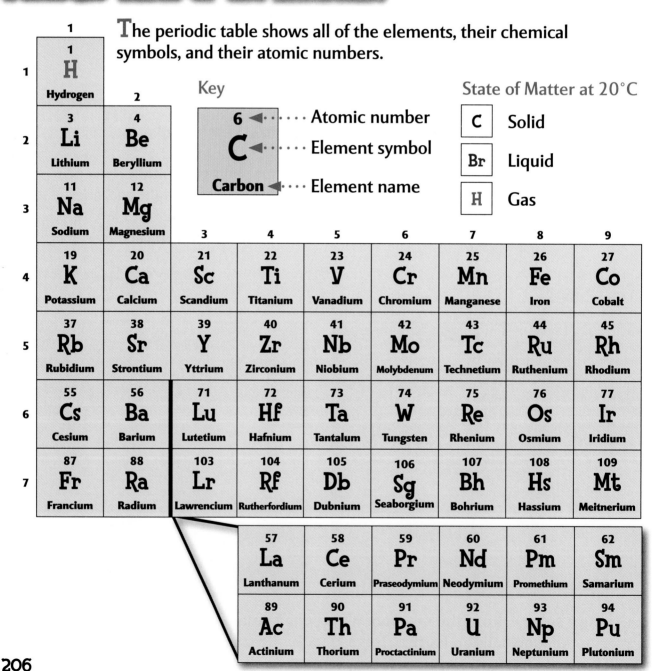

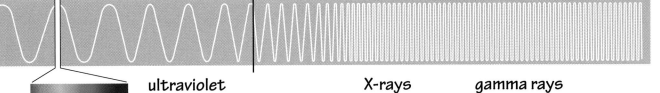

short wavelength (less than 0.000 000 1 mm)

ultraviolet   X-rays   gamma rays

Visible light

Metallic Properties

Li	Metal
B	Metalloid
C	Nonmetal

	10	11	12	13	14	15	16	17	18
									2 **He** Helium
				5 **B** Boron	6 **C** Carbon	7 **N** Nitrogen	8 **O** Oxygen	9 **F** Fluorine	10 **Ne** Neon
				13 **Al** Aluminum	14 **Si** Silicon	15 **P** Phosphorus	16 **S** Sulfur	17 **Cl** Chlorine	18 **Ar** Argon
	28 **Ni** Nickel	29 **Cu** Copper	30 **Zn** Zinc	31 **Ga** Gallium	32 **Ge** Germanium	33 **As** Arsenic	34 **Se** Selenium	35 **Br** Bromine	36 **Kr** Krypton
	46 **Pd** Palladium	47 **Ag** Silver	48 **Cd** Cadmium	49 **In** Indium	50 **Sn** Tin	51 **Sb** Antimony	52 **Te** Tellurium	53 **I** Iodine	54 **Xe** Xenon
	78 **Pt** Platinum	79 **Au** Gold	80 **Hg** Mercury	81 **Tl** Thallium	82 **Pb** Lead	83 **Bi** Bismuth	84 **Po** Polonium	85 **At** Astatine	86 **Rn** Radon
	110 **Ds** Darmstadtium	111 **Rg** Roentgenium	112	113	114	115	116		118

63 **Eu** Europium	64 **Gd** Gadolinium	65 **Tb** Terbium	66 **Dy** Dysprosium	67 **Ho** Holmium	68 **Er** Erbium	69 **Tm** Thulium	70 **Yb** Ytterbium
95 **Am** Americium	96 **Cm** Curium	97 **Bk** Berkelium	98 **Cf** Californium	99 **Es** Einsteinium	100 **Fm** Fermium	101 **Md** Mendelevium	102 **No** Nobelium

Glossary

To hear the glossary in English or Spanish, listen to the Video Glossary on one of the DVDs.

Webster here. Let me show you how to use the Glossary!

Science word

How you say the word

Science definition

leaf (leef) a plant part that grows from the stem and helps the plant get air and make food (94)

*Green is the most common **leaf** color.*

How the word is used in a sentence

Look on this page to see the word in the book.

adaptation (ad uhp TAY shuhn) a special characteristic that helps an organism to survive (38)

*A giraffe's long neck is an **adaptation**.*

air mass (ayr mas) a large region of the atmosphere where the air has similar characteristics (90)

*The warm **air mass** brought higher temperatures as it moved over our area.*

air pressure (ayr PRESH ur) the force put on a given area by the weight of the air above it (90)

*When you go up in an airplane, you can feel the difference in **air pressure**.*

anemometer (an uh MOM i tur) a tool that measures wind speed (104)

*An **anemometer** spins like a pinwheel.*

asteroid (AS tur oyd) a chunk of rock and metal that orbits the sun (119)

*The **asteroid** belt is located between Mars and Jupiter.*

Glossary

atmosphere (AT mus feer) the gases or air that surrounds Earth (90)

We get oxygen from the atmosphere.

attract (ah TRAKT) to pull toward itself (142)

A magnet will attract metal objects.

axis (AK sis) the imaginary line through the center of Earth (112)

Earth is tilted on its axis, which causes the seasons.

backbone (BAK bohn) a long line of bones that runs down the back of some animals (8)

A snake has a backbone.

bacteria (BAK teer ee uh) one-celled living things (54)

Bacteria are very small.

barometer (buh ROM i tur) a tool that measures air pressure (104)

The barometer showed that the air pressure was high.

battery (BAT ur ee) a source of electric current (186)

The batteries in the flashlight were running low.

camouflage (KAM uh flahzh) an adaptation that allows animals to blend into their surroundings (38)

A stick insect uses camouflage.

carbohydrates (kahr boh HEYE drayts) nutrients found in food from plants that give your body energy (46)

Athletes eat a lot of carbohydrates for energy.

chemical change (KEM I kuhl chaynj) a change that forms a different kind of matter (162)

When wood is burning, it goes through a chemical change.

Glossary

circuit (SUR kit) a complete path through which electricity can flow (186)

*The **circuit** must be complete to turn on the light.*

clay (klay) a kind of soil that can be molded when wet and that turns hard when dried (77)

*Pots can be made from **clay**.*

community (kuh MYEW ni tee) all the living things in an ecosystem (24)

*Your neighborhood is a **community**.*

condensation (kon den SAY shuhn) when water cools and changes from a gas into a liquid (98)

***Condensation** forms on a glass filled with ice water.*

conductor (kuhn DUK tur) a material that heat travels through easily (170)

*Metal is a good **conductor**.*

conserve (kon SURV) to save, protect, or use something wisely without wasting it (60, 192)

*It is important to **conserve** energy.*

constellation (kahn stuh LAY shuhn) a group of stars that appears to form a pattern (126)

*The **constellation** Orion is easy to find in the night sky.*

decomposer (dee cuhm POH zur) an organism that breaks down dead plant and animal material (30)

*A mushroom is a **decomposer**.*

desert (DEZ urt) an area that gets little rainfall, making it very dry (38)

*A cactus grows in the **desert**.*

disease (di ZEEZ) an illness or sickness of a plant or animal (54)

*Germs can cause **disease**.*

Glossary

dwarf planet (dwawrf PLAN it) a round, or nearly round, body that revolves around the sun and has not cleared a path within its own orbit (121)

*Pluto is a **dwarf planet**.*

earthquake (URTH kwayk) a sudden movement in the rocks that make up Earth's crust (68)

*The ground moves during an **earthquake**.*

ecosystem (EE koh sis tuhm) all the living and nonliving things in an environment and all their interactions (24)

*Everything around you is part of your **ecosystem**.*

electric current (ee LEK trik KUR uhnt) electricity that flows through a circuit (142)

*A strong **electric current** flows through power lines.*

electromagnet (ee lek troh MAG net) a temporary magnet created when current flows through wire wrapped around an iron bar (142)

*An **electromagnet** can be used to run a motor.*

endangered (EN dayn jurd) close to becoming extinct; having very few of its kind left (60)

*Elephants are **endangered** animals.*

energy (EN uhr jee) the ability to do work (192)

*Walking up a hill uses a lot of **energy**.*

environment (ehn VEYE ruhn muhnt) the things that make up an area, such as land, water, and air (2)

*Recycling is good for the **environment**.*

erosion (ee RO zhuhn) the carrying away of weathered materials (68)

***Erosion** happens when rivers carry weathered materials downstream.*

evaporation (i vap uh RAY shuhn) when water warms and changes from a liquid into a gas (98)

*A puddle dries up because of **evaporation**.*

extinct (EKS tinkt) died out; as when a species no longer exists (60)

*Every dinosaur is now **extinct**.*

Glossary

first quarter (fuhrst KWAWR tuhr) the phase of the moon in which the right half is visible and growing larger (115)

*The **first quarter** moon looked like a half-circle in the sky.*

flowering plant (FLOW uhr ing plant) a plant that produces seeds inside of flowers (8)

*A clover is a **flowering plant.***

food chain (fewd chayn) a path that energy follows as it moves from one type of animal to another (30)

*Humans are at the top of the **food chain.***

food web (fewd web) several food chains that are connected (30)

*A **food web** can be very complex.*

force (fawrs) a push or pull (134)

*He used **force** to push the box.*

forest (FAWR ist) land area with a heavy growth of trees (24)

*The **forest** was home to many different types of birds.*

fossil (FOS uhl) The preserved remains or trace of an organism that lived long ago (74)

***Fossils** of dinosaur footprints show that they traveled in herds.*

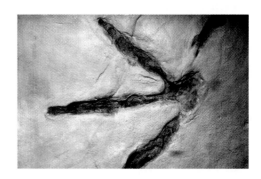

fossil fuel (FOS uhl fyewl) a substance, such as coal or oil, that was formed millions of years ago from the remains of plants and animals (82)

***Fossil fuels** are found underground.*

freeze (freez) to turn from a liquid into a solid (170)

*Extreme cold makes water **freeze.***

friction (FRIK shuhn) a force that occurs when one object rubs against another (134)

*Using sandpaper causes **friction.***

Glossary

front (frunt) a boundary between air masses that have different temperatures (90)

The cold front was marked in blue on the weather map.

fruit (frewt) the part of a plant that grows around seeds (18)

A watermelon is a large fruit.

fuel (fyewl) a substance burned for its energy (192)

Coal is an important fuel.

full moon (fuhl mewn) the phase of the moon in which all of its sunlit half is visible from Earth (115)

The full moon was a bright circle in the night sky.

fungus (FUHNG guhs) a one- or many-celled organism that absorbs food from other organisms (54)

Bread mold is a type of fungus.

gas (gas) matter that has no definite shape or volume (162)

Water vapor is a gas.

germinate (JUHR min ayt) to begin to grow, as when the right conditions allow a seed to develop (19)

Flower seeds germinate in the spring.

glacier (GLAY shur) a large mass of moving ice (68)

Glaciers were common during the ice ages.

gravity (GRAV i tee) a pulling force between two objects, such as Earth and you (112)

Earth's gravity keeps the moon in orbit.

habitat (HAB uh tat) the home of a living thing (24)

A tree is a squirrel's habitat.

heat (heet) the flow of energy from warmer matter to cooler matter (170)

A fire gives off heat to the air around it.

hibernate (HEYE buhr nayt) to rest or sleep through the cold winter (38)

Bears hibernate in caves or dens.

humus (HYEW muhs) decayed plant or animal material in soil (74)

The dead leaves will form humus.

Glossary

hurricane (HUR i kayn) a huge storm that forms over warm ocean water (104)

*The **hurricane** caused a lot of damage.*

igneous rock (IG nee uhs rahk) a fire-made rock formed from melted rock material (75)

*Lava cools to form **igneous rock.***

inclined plane (in KLINED playn) a flat surface that is raised at one end (148)

*A ramp is an example of an **inclined plane.***

inexhaustible resource (in ek ZAWS ti buhl REE sawrs) one that cannot be used up easily (82)

*Solar energy is an **inexhaustible resource.***

inner planet (IN uhr PLAN it) any of the four planets in the solar system that are closest to the sun (120)

*Mercury, Venus, Earth, and Mars are **inner planets.***

insect (IN sekt) an animal with three body sections, three pairs of legs, and, usually, two pairs of wings (8)

***Insects** are the largest group of animals in the world.*

insulator (IN suh lay tur) a material that heat doesn't travel through easily (170)

*Wood is a good **insulator.***

invertebrate (in VUR tuh brit) an animal that does not have a backbone (10)

*An earthworm is an **invertebrate.***

landform (LAND fawrmz) a natural feature on Earth's surface (68)

*Mount Everest is a famous **landform.***

leaf (leef) a plant part that grows from the stem and helps the plant get air and make food (2)

*Green is the most common **leaf** color.*

Glossary

lever (LEV ur) a straight bar that moves on a fixed point (148)

*A seesaw is an example of a **lever**.*

life cycle (leyef SEYE kel) all the stages in an organism's life (18)

*A frog has several stages in its **life cycle**.*

liquid (LIK wid) matter that has a definite volume but not a definite shape (162)

***Liquid** water takes the shape of whatever container it is in.*

magnetism (MAG ni tiz uhm) a force that exists around a magnet and exerts a pull on most metals, such as iron and steel (142)

***Magnetism** plays an important role in many machines.*

mammal (MAM uhl) an animal with hair or fur that feeds its young with milk (8)

*A human being is a **mammal**.*

mass (mas) the measure of the amount of matter in an object (156)

*A heavy object has a large **mass**.*

matter (MAT ur) anything that takes up space and has mass (156)

*Air is **matter**, even though you can't see it.*

melt (melt) to change from a solid to a liquid (170)

*Ice cream will **melt** if you don't keep it cold.*

metamorphic rock (met uh MOR fik rahk) a rock that has changed form through squeezing and heating (75)

***Metamorphic rock** forms deep within Earth.*

metamorphosis (met uh MAWR fuh sis) a change in the body form of an organism (18)

*A caterpillar goes through **metamorphosis** to change into a butterfly.*

Glossary

meteorologist (mee tee awr OL oh jist) a person who studies the weather (104)

The weatherperson on the news is a meteorologist.

microorganism (MEYE kroh OR guh ni zum) an organism that is so small you need a microscope to see it (54)

There are microorganisms around you all the time.

microscope (MEYE kruh skohp) a device that uses glass lenses to allow people to see very small things (54)

You can use a microscope to study cells.

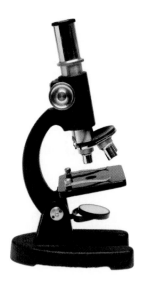

migrate (MEYE grayt) to move to another place (38)

Many birds migrate south in the winter.

mimicry (MIM i kree) the imitation by one animal of the traits of another (38)

Some animals use mimicry to confuse predators.

mineral (MIN ur uhl) nonliving material that can be found in rocks and soil (74)

All rocks are made from minerals.

minerals (MIN uhr ulz) nutrients that work to help control body processes (46)

Your body needs vitamins and minerals.

mold (mohld) a type of fungus that grows in damp places (54)

Mold can grow on food.

motion (MOH shuhn) a change in position (134)

A runner is in motion during a race.

natural resource (NACH uhr uhl REE sawrs) a material on Earth that is necessary or useful to people (82)

Soil is an important natural resource.

Glossary

new moon (new mewn) the phase of the moon in which none of its sunlit half is visible from Earth (115)

*The **new moon** cannot be seen in the sky.*

nonrenewable resource (non ri NEW uh buhl REE sawrs) one that cannot be reused or replaced easily (82)

*Fossil fuels are **nonrenewable resources.***

nutrients (NEW tree uhnts) substances that provide energy and control body processes (46)

*Our bodies get **nutrients** from food.*

orbit (AWR bit) the path an object follows as it revolves around another object (118)

*Earth is in **orbit** around the sun.*

organism (AWR guh niz uhm) any living thing (2)

***Organisms** can be many different shapes and sizes.*

outer planet (OWT uhr PLAN it) any of the four planets in the solar system that are farthest from the sun (121)

*Jupiter, Saturn, Uranus, and Neptune are the **outer planets.***

oxygen (OX suh juhn) a gas that is in air and water (2)

*People need **oxygen** to live.*

phase (fayz) the shape of the lighted part of the moon as seen from Earth (112)

*The **phase** of the moon changes from day to day.*

Glossary

physical change (FIZ i kuhl chaynj) a change in the way matter looks that leaves the matter itself unchanged (162)

*Melting ice is going through a **physical change**.*

pitch (pich) how high or low a sound is (178)

*Dogs can hear sounds with a high **pitch**.*

planet (PLAN it) a large round, or nearly round, body that revolves around the sun and has cleared a path within its orbit (118)

*Earth is the only **planet** known to support life.*

poles (pohlz) two ends of a magnet where the magnetic force is strongest (144)

*The magnet's **poles** attracted the paper clips.*

pollen (POL uhn) a powdery material needed by the eggs of flowers to make seeds (18)

*Plants make **pollen** to reproduce.*

pollution (puh LEW shuhn) the adding of harmful substances to the water, air, or land (60)

***Pollution** is bad for the environment.*

population (pop yuh LAY shuhn) all the members of a single type of an organism in an ecosystem (24)

*There is a large **population** of pigeons in the city.*

position (puh ZISH uhn) the location of an object at a particular point in time (134)

*The player changed **position** during the game.*

Glossary

precipitation (pri sip i TAY shuhn) water in the atmosphere that falls to Earth as rain, snow, hail, or sleet (98)

*The weatherperson forecasted lots of **precipitation**.*

predator (PREHD uh tawr) an animal that hunts other animals for food (30)

*My cat is a **predator** of mice.*

prey (pray) the animals that predators eat (30)

*A cheetah chases its **prey**.*

prism (PRIZ uhm) a block of angled glass that bends the waves of light passing through it (178)

*The **prism** showed colors on the wall.*

property (PROP ur tee) something you can observe with your senses (156)

*The color of a rock is a **property**.*

protein (PROH teen) a substance found in plants and animals that keeps a body strong (46)

*Many people get **protein** from eating meat.*

pulley (PUL ee) a simple machine that uses a wheel and rope to lift a weight (148)

*The **pulley** made lifting the heavy box easier.*

r

reflect (ri FLEKT) to bounce light off of a surface (178)

*Shiny objects often **reflect** light.*

refract (ri FRAKT) to bend light as it passes through matter (180)

*Light can **refract** when it passes through water.*

renewable resource (ri NEW uh buhl REE sawrs) one that can be replaced in a short period of time (82)

*Wood is an example of a **renewable resource**.*

repel (ri PEL) to push away (142)

*Magnets can **repel** other magnets.*

reproduction (ree pruh DUK shuhn) the way organisms make more of their own kind (18)

***Reproduction** is important to the survival of a kind of organism.*

resource (REE sawrs) a material on Earth that is necessary or useful to living things (60)

*Food is an important **resource** for people.*

Glossary

revolve (ri VOLV) to move in a nearly circular path around something else (112)

*The planets **revolve** around the sun.*

rocket (RAH kit) a tube-shaped device driven through the air or space by a stream of hot gases (126)

*A **rocket** can be used in space exploration.*

root (root) a plant part that takes in water and minerals (2)

*Most **roots** grow underground.*

rotate (ROH tayt) to spin on an axis (112)

*A merry-go-round **rotates** when people spin it.*

sand (sand) tiny, loose grains of crushed rocks (77)

*The **sand** at a beach was formed from larger rocks.*

satellite (SAT uhl leyet) any object that orbits another larger body in space (126)

*The moon is Earth's only natural **satellite**.*

screw (screw) an inclined plane wrapped into a spiral (148)

*A **screw** can hold things together better than a nail.*

seasons (SEE zuhnz) spring, summer, fall, and winter (112)

*In many places, the four **seasons** have very different weather.*

sedimentary rock (sed uh MENT uh ree rahk) a kind of rock formed when sand, mud, or pebbles at the bottom of rivers, lakes, and oceans pile up (75)

*Sandstone is a common **sedimentary rock**.*

seed plant (seed plant) a plant that produces and grows from seeds (8)

*Fruit comes from **seed plants**.*

Glossary

seedling (SEED ling) a young plant (18)

*Huge trees grow from tiny **seedlings**.*

silt (silt) a kind of powdery, slippery soil (77)

***Silt** is often found at the bottom of rivers.*

simple machine (SIM puhl muh SHEEN) a machine with few or no moving parts. Simple machines include the wedge, lever, inclined plane, screw, pulley, and wheel and axle (149)

*A lever is an example of a **simple machine**.*

soil (soyl) a mix of tiny rock particles, minerals, and decayed plant and animal materials (74)

*Most plants need **soil** to grow.*

solar energy (SOH lur EN uhr jee) energy from the sun (192)

*Some machines can run on **solar energy**.*

solar system (SOH lur SIS tem) includes the sun and all the objects that orbit the sun (118)

*There are many planets, asteroids, and comets in our **solar system**.*

solid (SOL id) matter that has a definite shape and volume (162)

*Ice is a **solid**.*

space probes (spays prohbz) rocket-launched vehicles that carry data-gathering equipment into space (126)

*The **space probe** sent back new photos of Mars.*

space shuttle (spays SHUHT uhl) a reusable space vehicle that is launched into orbit like a rocket and returns to Earth to land like an airplane (129)

*Astronauts on board the **space shuttle** launched the Hubble telescope.*

star (stahr) a huge, hot sphere of gases that gives off its own light (118)

*The sun is an example of a **star**.*

state of matter (stayt uhv MAT ur) any of the forms that matter can take, including solid, liquid, or gas (163)

*Water can exist in any **state of matter**.*

stem (stehm) a plant part that supports the plant (2)

*An oak tree has a big **stem**.*

Glossary

switch (swich) a device that can open or close an electric circuit (186)

*Flipping the **switch** turned off the light.*

telescope (TEL uh skohp) a tool that gathers light to make faraway objects appear closer (126)

*Astronomers use **telescopes** to study space.*

texture (TEKS chur) how the surface of an object feels to the touch (156)

*A rock can have a rough **texture**.*

thermometer (thuhr MAHM i tuhr) a tool that measures the temperature (104)

*The **thermometer** shows a very low temperature in the winter.*

third quarter (thuhrd KWAWR tuhr) the phase of the moon in which the left half is visible and growing smaller. Also called the last quarter moon (115)

*The **third quarter** moon will soon become a new moon.*

tornado (tawr NAY doh) a violent, spinning wind that moves across the ground in a narrow path (104)

*A **tornado** can be very destructive.*

vertebrate (VUR tuh brit) an animal that has a backbone (8)

*All mammals are **vertebrates**.*

vibrate (VEYE brayt) to move back and forth quickly (178)

*The elephants' footsteps caused the ground to **vibrate**.*

vitamin (VEYE tah min) a nutrient in foods that the body needs for growth (46)

***Vitamin** C is found in orange juice.*

Glossary

volcano (vol KAY noh) an opening in the surface of Earth from which lava flows (68)

Each island in Hawaii was formed by a volcano.

volume (VOL yewm) the measure of how much space matter takes up; how loud or soft a sound is (156, 178)

The swimming pool has a large volume of water. The volume of the music was very loud.

water cycle (WAH tur SEYE kel) the constant movement of water from Earth to the atmosphere and back to Earth again (98)

Farmers rely on the water cycle for rain.

water vapor (WAH tur VAY puhr) water in the form of a gas (98)

There is a lot of water vapor in the air.

weather (WETH ur) the condition of the atmosphere at a given time and place (90)

The weather was stormy today.

weathering (WETH ur ing) the process that causes rocks to crumble, crack, and break (68)

Weathering breaks huge rocks down into tiny rocks.

wedge (WEJ) two inclined planes placed back-to-back (148)

A wedge can be used to split wood.

weight (wayt) the measure of the pull of gravity on an object (157)

Your weight on Earth is different from your weight on the moon.

wetland (WET land) land area that contains a lot of moisture (24)

A wetland is home to many animals.

wheel and axle (hweel and AK sul) a wheel that turns on a post (148)

A doorknob is a wheel and axle.

wind (wind) moving air (90)

The wind blew the leaves from the trees.

Glosario

a

adaptation/adaptación una característica especial que ayuda a un organismo a sobrevivir (38)

*El cuello largo de una jirafa es una **adaptación**.*

air mass/masa de aire una región grande de la atmósfera donde el aire tiene características similares (90)

*La **masa de aire** cálido trajo temperaturas más elevadas a medida que avanzó sobre nuestra área.*

air pressure/presión del aire la fuerza ejercida por el peso del aire sobre un área dada (90)

*Cuando viajas en un avión, puedes sentir la diferencia en la **presión del aire**.*

anemometer/anemómetro un instrumento que mide la velocidad del viento (104)

*Un **anemómetro** gira como una veleta.*

asteroid/asteroide un pedazo de roca y metal que orbita alrededor del Sol (119)

*El cinturón de **asteroides** está ubicado entre Marte y Júpiter.*

atmosphere/atmósfera los gases o el aire que rodean la Tierra (90)

*El oxígeno se obtiene de la **atmósfera**.*

attract/atraer jalar hacia sí mismo (142)

*Un imán **atrae** objetos de metal.*

axis/eje la línea imaginaria que atraviesa el centro de la Tierra (112)

*La Tierra está inclinada sobre su **eje**, lo cual origina las estaciones.*

b

backbone/columna vertebral una línea larga de huesos que se extiende a lo largo de la espalda o lomo de algunos animales (8)

*La serpiente tiene **columna vertebral**.*

bacteria/bacterias seres vivos unicelulares (54)

*Las **bacterias** son muy pequeñas.*

barometer/barómetro un instrumento que mide la presión del aire (104)

*El **barómetro** indicó que la presión del aire era elevada.*

Glosario

battery/pila una fuente de corriente eléctrica (186)

*Se estaban acabando las **pilas** de la linterna.*

camouflage/camuflaje una adaptación que permite a los animales confundirse con el ambiente que los rodea (38)

*Un insecto palo usa el **camuflaje**.*

carbohydrates/ carbohidratos nutrientes que se encuentran en los alimentos provenientes de las plantas, que dan energía a tu cuerpo (46)

*Los atletas consumen muchos **carbohidratos** como fuente de energía.*

chemical change/cambio químico un cambio que produce una transformación de la materia (162)

*Cuando la madera se quema, atraviesa un **cambio químico**.*

circuit/circuito una trayectoria completa a través de la cual se puede conducir la electricidad (186)

*El **circuito** debe ser completo para encender la luz.*

clay/arcilla un tipo de suelo que puede moldearse cuando está húmedo y se endurece cuando se seca (77)

*Las vasijas se pueden hacer con **arcilla**.*

community/comunidad todos los seres vivos de un ecosistema (24)

*Tu vecindario es una **comunidad**.*

condensation/condensación cuando el agua se enfría y se transforma de un gas a un líquido (98)

*En un vaso lleno de agua helada se forma **condensación**.*

conductor/conductor un material por el cual el calor viaja fácilmente (170)

*El metal es un buen **conductor**.*

conserve/conservar ahorrar, proteger o usar algo prudentemente sin desperdiciarlo (60, 192)

*Es importante **conservar** la energía.*

Glosario

constellation/constelación un grupo de estrellas que aparentemente forman un patrón (126)

*Es fácil ubicar la **constelación** de Orión en el cielo nocturno.*

decomposer/descomponedor un organismo que descompone desechos de plantas y animales muertos (30)

*Un hongo es un **descomponedor**.*

desert/desierto un área que recibe muy poca lluvia y, por lo tanto, es muy seca (38)

*Los cactus crecen en el **desierto**.*

disease/enfermedad afección de una planta o animal (54)

*Los gérmenes pueden causar **enfermedades**.*

dwarf planet/planeta enano un cuerpo redondo o casi redondo que gira alrededor del Sol y no ha despejado una trayectoria dentro de su propia órbita (121)

*Plutón es **planeta enano**.*

earthquake/terremoto un movimiento repentino de las rocas que componen la corteza terrestre (68)

*Durante un **terremoto**, la tierra se mueve.*

ecosystem/ecosistema todas los seres vivos y sin vida de un medio ambiente y todas sus interacciones (24)

*Todo lo que te rodea forma parte de tu **ecosistema**.*

electric current/corriente eléctrica electricidad que se conduce a través de un circuito (142)

*Una fuerte **corriente eléctrica** se conduce a través de los cables de alta tensión.*

electromagnet/electroimán un imán creado temporalmente cuando la corriente se conduce a través de un alambre enroscado a una barra de hierro (142)

*Un **electroimán** se puede usar para hacer funcionar un motor.*

Glosario

endangered/en peligro de extinción
que tiene un gran riesgo de
extinguirse; que quedan muy pocos
de su especie (60)

*Los elefantes están **en peligro de
extinción**.*

energy/energía la capacidad de
realizar un trabajo (192)

*Para subir una colina se requiere
mucha **energía**.*

environment/medio ambiente todo
aquello que compone un área, tal
como la tierra, el agua y el aire (2)

*Reciclar es bueno para el **medio
ambiente**.*

erosion/erosión el desplazamiento de
materiales desgastados (68)

*La **erosión** ocurre cuando los ríos
arrastran materiales desgastados.*

evaporation/evaporación cuando el
agua se calienta y se transforma de
un líquido a un gas (98)

*Los charcos se secan debido a la
evaporación.*

extinct/extinto desaparecido; como
cuando una especie no existe más
(60)

*Todos los dinosaurios actualmente
están **extintos**.*

first quarter/cuarto creciente la
fase de la Luna en la cual se ve la
mitad derecha, en su período de
crecimiento (115)

*La Luna en **cuarto creciente** parecía
media esfera en el cielo.*

flowering plant/planta con flores
una planta que produce semillas
dentro de las flores (8)

*El trébol es una **planta con flores**.*

food chain/cadena alimenticia
un camino que la energía sigue a
medida que se mueve de un tipo de
animal a otro (30)

*Los seres humanos están en la parte
más alta de la **cadena alimenticia**.*

food web/red alimenticia varias
cadenas alimenticias que están
conectadas (30)

*Una **red alimenticia** puede ser muy
compleja.*

Glosario

force/fuerza un empujón o jalón (134)

*Él usó la **fuerza** para empujar la caja.*

forest/bosque extensión de tierra donde crecer muchos árboles (24)

*El **bosque** fue el hogar de muchos tipos de aves.*

fossil/fósil los restos o rastros conservados de un organismo que vivió hace mucho tiempo (74)

*Los **fósiles** de huellas de dinosaurios demuestran que viajaban en manadas.*

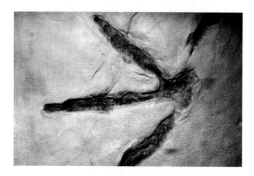

fossil fuel/combustible fósil una sustancia, como el carbón o el petróleo, que se formó hace millones de años a partir de restos de plantas y animales (82)

*Los **combustibles fósiles** se encuentran bajo tierra.*

freeze/congelar convertir un líquido en sólido (170)

*El frío extremo hace que el agua se **congele**.*

friction/fricción una fuerza que se produce cuando un objeto se frota con otro (134)

*Al usar una lija se produce **fricción**.*

front/frente un límite entre masas de aire que tienen distintas temperaturas (90)

*El **frente** frío estaba señalado en azul en el mapa meteorológico.*

fruit/fruta parte de la planta que crece alrededor de las semillas (18)

*La sandía es una **fruta** grande.*

fuel/combustible una sustancia que se quema para obtener energía (192)

*El carbón es un **combustible** importante.*

full moon/luna llena la fase de la Luna en la cual la mitad que está iluminada por el Sol es visible desde la Tierra (115)

*La **luna llena** era una esfera luminosa en el cielo nocturno.*

fungus/hongo un organismo unicelular o pluricelular que absorbe alimento de otros organismos (54)

*El moho del pan es un tipo de **hongo**.*

Glosario

g

gas/gas materia que no tiene ni forma ni volumen definidos (162)

El vapor de agua es un gas.

germinate/germinar empezar a crecer, como cuando se dan las condiciones apropiadas que le permiten a la semilla desarrollarse (19)

Las semillas de las flores germinan en la primavera.

glacier/glaciar una masa grande de hielo en movimiento (68)

Los glaciares eran comunes durante las edades de hielo.

gravity/gravedad una fuerza de atracción entre dos objetos, como entre la Tierra y tú (112)

La gravedad de la Tierra mantiene a la Luna en órbita.

h

habitat/hábitat el hogar de un ser vivo (24)

Un árbol es el hábitat de una ardilla.

heat/calor el flujo de energía de la materia más caliente a la materia más fría (170)

Un incendio despide calor al aire circundante.

hibernate/hibernar descansar o dormir durante el frío invierno (38)

Los osos hibernan en cuevas o madrigueras.

humus/humus material vegetal o animal en descomposición que se halla en el suelo (74)

Las hojas muertas formarán humus.

hurricane/huracán una gran tormenta que se forma sobre agua cálida del océano (104)

El huracán provocó grandes daños.

i

igneous rock/roca ígnea una roca "hecha a partir del fuego" que se formó a partir de material rocoso fundido (75)

Al enfriarse la lava, se forma la roca ígnea.

inclined plane/plano inclinado una superficie plana elevada en uno de sus extremos (148)

Una rampa es un ejemplo de un plano inclinado.

Glosario

inexhaustible resource/
recurso inagotable recurso que
no puede agotarse fácilmente (82)

*La energía solar es un **recurso
inagotable**.*

inner planet/planeta
interior cualquiera de los cuatro
planetas del Sistema Solar que están
más cerca del Sol. (120)

*Mercurio, Venus, la Tierra y Marte son
planetas interiores.*

insect/insecto un animal cuyo cuerpo
está compuesto por tres secciones,
tres pares de patas y, normalmente,
dos pares de alas (8)

*Los **insectos** son el grupo de animales
más grande del mundo.*

insulator/aislante un material
en el cual el calor no viaja
fácilmente (170)

*La madera es un buen **aislante**.*

invertebrate/invertebrado un animal
que no tiene columna vertebral (10)

*Un lombriz de tierra es un
invertebrado.*

landform/accidente geográfico
una característica natural en la
superficie terrestre (68)

*El monte Everest es un **accidente
geográfico** conocido.*

leaf/hoja una parte de la planta que
crece desde el tallo y le ayuda a la
planta a obtener aire y procesar
alimento (2)

*El color más común de una **hoja** es
el verde.*

lever/palanca una barra recta que se
mueve sobre un punto fijo (148)

*Un subibaja es un ejemplo de **palanca**.*

life cycle/ciclo de vida todas las etapas
de la vida de un organismo (18)

*El **ciclo de vida** de una rana tiene
varias etapas.*

liquid/líquido materia que tiene un
volumen definido pero no una
forma definida (162)

*El agua **líquida** toma la forma del
recipiente en que se encuentra.*

Glosario

magnetism/magnetismo una fuerza que existe alrededor de un imán y atrae a la mayoría de los metales, como el hierro y el acero (142)

*El **magnetismo** cumple una función importante en muchas máquinas.*

mammal/mamífero un animal que tiene pelo o pelaje y que amamanta a sus crías (8)

*El ser humano es un **mamífero**.*

mass/masa la medida de la cantidad de materia en un objeto (156)

*Los objetos pesados tienen una **masa** grande.*

matter/materia todo lo que ocupa un lugar en el espacio y tiene masa (156)

*El aire es **materia**, aunque no lo podamos ver.*

melt/derretir convertir un sólido en líquido (170)

*El helado se **derrite** si no lo mantienes frío.*

metamorphic rock/ roca metamórfica una roca que ha cambiado su forma por medio del calor y la presión (75)

*La **roca metamórfica** se forma en las profundidades de la Tierra.*

metamorphosis/metamorfosis un cambio que se produce en la forma del cuerpo de un organismo (18)

*La oruga pasa por la **metamorfosis** para convertirse en mariposa.*

meteorologist/meteorólogo una persona que estudia el tiempo (104)

*El experto del tiempo en el noticiero es un **meteorólogo**.*

Glosario

microorganism/microorganismo
un organismo tan pequeño que
se necesita un microscopio para
observarlo (54)

*Hay **microorganismos** a tu alrededor
todo el tiempo.*

microscope/microscopio
un instrumento que permite a las
personas ver cosas muy pequeñas a
través de lentes de vidrio (54)

*Puedes usar un **microscopio** para
estudiar las células.*

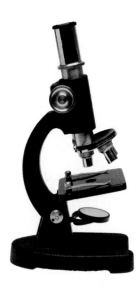

migrate/migrar trasladarse a otro
lugar (38)

*Muchas aves **migran** hacia el sur en
el invierno.*

mimicry/mimetismo la imitación por
parte de un animal de los rasgos de
otro (38)

*Algunos animales usan el **mimetismo**
para confundir a los depredadores.*

mineral/mineral material sin vida que
se encuentra en las rocas y en el
suelo (74)

*Todas las rocas están hechas de
minerales.*

minerals/minerales nutrientes que
sirven para ayudar a controlar el
funcionamiento del cuerpo (46)

*Tu cuerpo necesita vitaminas y
minerales.*

mold/moho un tipo de hongo que
crece en lugares húmedos (54)

*El **moho** puede crecer en los alimentos.*

motion/movimiento un cambio de
posición (134)

*Un corredor está en **movimiento**
durante una carrera.*

natural resource/recurso natural
un material de la Tierra que
es necesario o útil para las
personas (82)

*El suelo es un **recurso natural**
importante.*

new moon/luna nueva la fase de la
Luna en la cual la mitad que está
iluminada por el Sol no es visible
desde la Tierra (115)

*La **luna nueva** no se ve en el cielo.*

Glosario

nonrenewable resource/ recurso no renovable recurso que no puede volver a usarse ni reemplazarse fácilmente (82)

Los combustibles fósiles son recursos no renovables.

nutrients/nutrientes sustancias que proveen energía y controlan el funcionamiento del cuerpo (46)

Nuestro cuerpo recibe nutrientes de los alimentos.

orbit/órbita la trayectoria que sigue un objeto mientras gira alrededor de otro objeto (118)

La Tierra está en órbita alrededor del Sol.

organism/organismo cualquier ser vivo (2)

La forma y el tamaño de los organismos pueden ser muy diferentes.

outer planet/planeta exterior cualquiera de los cuatro planetas del Sistema Solar que están más lejos del Sol. (121)

Los planetas exteriores son Júpiter, Saturno, Urano, Neptuno y Plutón.

oxygen/oxígeno un gas que se encuentra en el aire y el agua (2)

Las personas necesitan oxígeno para vivir.

phase/fase la forma de la parte iluminada de la Luna vista desde la Tierra (112)

La fase de la Luna cambia día a día.

physical change/cambio físico un cambio en la forma en que se presenta la materia sin modificar la materia misma (162)

El hielo que se derrite atraviesa un cambio físico.

pitch/tono qué tan alto o bajo es un sonido (178)

Los perros pueden escuchar sonidos de tono alto.

Glosario

planet/planeta un cuerpo redondo o casi redondo, de gran tamaño, que gira alrededor del Sol y ha despejado una trayectoria dentro de su propia órbita (118)

*La Tierra es el único **planeta** donde se sabe que hay vida.*

poles/polos los dos extremos de un imán en donde la fuerza magnética es más fuerte (144)

*Los **polos** del imán atrajeron los clips.*

pollen/polen un material parecido al polvo que necesitan los óvulos de las flores (también llamados gametos femeninos) para hacer semillas (18)

*Las plantas producen **polen** para reproducirse.*

pollution/contaminación el agregado de sustancias tóxicas al agua, el aire o la tierra (60)

*La **contaminación** es nociva para el medio ambiente.*

population/población todos los miembros de un mismo tipo de organismo en un ecosistema (24)

*Hay una gran **población** de palomas en la ciudad.*

position/posición la ubicación de un objeto en un momento dado (134)

*El jugador cambió de **posición** durante el juego.*

precipitation/precipitación el agua de la atmósfera que cae a la Tierra en forma de lluvia, nieve, granizo o aguanieve (98)

*El experto del tiempo pronosticó fuertes **precipitaciones**.*

predator/depredador un animal que caza otros animales para alimentarse (30)

*Mi gato es un **depredador** de ratones.*

prey/presa los animales que los depredadores se comen (30)

*Un guepardo persigue a su **presa**.*

prism/prisma un bloque de vidrio angulado que curva las ondas de luz que lo atraviesan (178)

*El **prisma** mostró colores en el muro.*

property/propiedad algo que puedes observar con los sentidos (156)

*El color de una roca es una **propiedad**.*

Glosario

protein/proteína una sustancia que se encuentra en las plantas y los animales y mantiene nuestro cuerpo saludable (46)

*Muchas personas obtienen **proteínas** al consumir carne.*

pulley/polea una máquina simple que utiliza una rueda y una cuerda para levantar peso (148)

*Gracias a la **polea,** fue más fácil levantar la caja.*

reflect/reflejar rebotar la luz en una superficie (178)

*Los objetos brillantes a menudo **reflejan** luz.*

refract/refractar curvar la luz a medida que atraviesa la materia (180)

*La luz puede **refractarse** cuando atraviesa el agua.*

renewable resource/ recurso renovable recurso que puede reemplazarse en un período corto de tiempo (82)

*La madera es un ejemplo de **recurso renovable.***

repel/repeler alejar (142)

*Los imanes pueden **repeler** otros imanes.*

reproduction/reproducción la forma en que los organismos se reproducen (18)

*La **reproducción** es importante para la supervivencia de un tipo de organismo.*

resource/recurso un material de la Tierra que es necesario o útil para los seres vivos (60)

*El alimento es un **recurso** importante para las personas.*

revolve/girar moverse en una trayectoria casi circular alrededor de otra cosa (112)

*Los planetas **giran** alrededor del Sol.*

rocket/cohete un aparato en forma de tubo que viaja por el aire o el espacio impulsado por una corriente de gases calientes (126)

*Los **cohetes** se usan en la exploración espacial.*

Glosario

root/raíz una parte de la planta que absorbe agua y minerales (2)

La mayoría de la raíces crecen bajo tierra.

rotate/rotar girar alrededor de un eje (112)

Un carrusel rota cuando se le hace girar.

sand/arena pequeños granos sueltos de roca pulverizada (77)

La arena en una playa se formó de rocas más grandes.

satellite/satélite cualquier objeto que orbita alrededor de otro cuerpo más grande en el espacio (126)

La Luna es el único satélite natural de la Tierra.

screw/tornillo un plano inclinado en forma de espiral (148)

Un tornillo puede mantener cosas unidas mejor que un clavo.

seasons/estaciones primavera, verano, otoño e invierno (112)

El tiempo de muchos lugares es muy diferente en cada una de las cuatro estaciones.

sedimentary rock/roca sedimentaria un tipo de roca que se forma cuando se acumula arena, lodo o guijarros en el fondo de ríos, lagos y océanos (75)

La arenisca es una roca sedimentaria común.

seed plant/planta con semillas una planta que produce semillas y crece a partir de ellas (8)

Las frutas crecen de las plantas con semillas.

seedling/plántula una planta joven (18)

Los árboles gigantescos crecen de pequeñísimas plántulas.

silt/cieno un tipo de suelo parecido al polvo y resbaloso (77)

El cieno se suele encontrar en el fondo de los ríos.

Glosario

simple machine/máquina simple
una máquina con pocas o ninguna parte movible. Entre ellas están la cuña, la palanca, el plano inclinado, el tornillo, la polea y la rueda y eje. (149)

Una palanca es un ejemplo de **máquina simple.**

soil/suelo una mezcla de pequeñas partículas de roca, minerales y materiales animales y vegetales en descomposición (74)

La mayoría de plantas necesitan **suelo** *para crecer.*

solar energy/energía solar energía del Sol (192)

Algunas máquinas pueden funcionar con **energía solar.**

solar system/Sistema Solar incluye el Sol y todos los objetos que orbitan alrededor del Sol (118)

Hay muchos planetas, asteroides y cometas en nuestro **Sistema Solar.**

solid/sólido materia que tiene forma y volumen definidos (162)

El hielo es un **sólido.**

space probes/sondas espaciales vehículos impulsados por cohetes que transportan al espacio equipos para la recolección de información (126)

La **sonda espacial** *transmitió nuevas fotos de Marte.*

space shuttle/trasbordador espacial un vehículo espacial reutilizable que se pone en órbita como un cohete y regresa a la Tierra y aterriza como un avión (129)

Los astronautas que estaban a bordo del **trasbordador espacial** *lanzaron el telescopio Hubble.*

star/estrella una inmensa esfera de gases calientes que emite su propia luz (118)

El Sol es un ejemplo de **estrella.**

state of matter/estado de la materia cualquiera de las formas que puede tomar la materia, ya sea sólida, líquida o gaseosa (163)

El agua puede existir en cualquier **estado de la materia.**

Glosario

stem/tallo una parte de la planta que la sostiene (2)

*Un roble tiene un **tallo** grande.*

switch/interruptor un dispositivo que abre o cierra un circuito eléctrico (186)

*Al manipular el **interruptor** se apagó la luz.*

telescope/telescopio un instrumento que reúne luz para hacer que los objetos lejanos aparezcan más cercanos (126)

*Los astrónomos usan **telescopios** para estudiar el espacio.*

texture/textura la manera en que se siente al tacto la superficie de un objeto (156)

*Una roca puede tener una **textura** áspera.*

thermometer/termómetro un instrumento que mide la temperatura (104)

*El **termómetro** indica una temperatura muy baja en el invierno.*

third quarter/cuarto menguante La fase de la Luna en la cual se ve la mitad izquierda, en su período decreciente. También se le llama el último cuarto de la Luna. (115)

*La Luna en **cuarto menguante** pronto se convertirá en luna nueva.*

tornado/tornado un viento giratorio violento que se desplaza por el suelo en una trayectoria estrecha (104)

*Un **tornado** puede ser muy destructivo.*

Glosario

vertebrate/vertebrado un animal que tiene columna vertebral (8)

Todos los mamíferos son vertebrados.

vibrate/vibrar moverse rápidamente hacia delante y hacia atrás (178)

Las pisadas de los elefantes hicieron que la tierra vibrara.

vitamin/vitamina un nutriente que se encuentra en los alimentos y es necesario para el crecimiento del cuerpo (46)

La vitamina C se encuentra en el jugo de naranja.

volcano/volcán una abertura en la superficie terrestre de la que brota lava (68)

Todas las islas de Hawai se formaron por volcanes.

volume/volumen la medida del espacio que ocupa la materia; qué tan fuerte o bajo es un sonido (156)

La piscina tiene un gran volumen de agua. El volumen de la música estaba muy fuerte.

water cycle/ciclo del agua el movimiento continuo del agua de la Tierra a la atmósfera y nuevamente de regreso a Tierra (98)

Los agricultores dependen del ciclo del agua para obtener lluvia para sus cultivos.

water vapor/vapor de agua agua en forma de gas (98)

Hay mucho vapor de agua en el aire.

weather/tiempo el estado de la atmósfera en un momento y lugar dados (90)

Hoy el tiempo estuvo tormentoso.

weathering/meteorización el proceso que hace las rocas de desmoronen, se agrieten y se rompan (68)

La meteorización convierte rocas enormes en rocas pequeñas.

wedge/cuña dos planos inclinados colocados uno contra el otro (148)

Una cuña puede usarse para separar madera.

Glosario

weight/peso la medida de la atracción de la gravedad sobre un objeto (157)

*Tu **peso** es diferente en la Tierra y en la Luna*

wetland/pantano extensión de tierra que contiene mucha humedad (24)

*Un **pantano** es el hogar de muchos animales.*

wheel and axle/rueda y eje una rueda que gira en un poste (148)

*Una manija es una **rueda y eje**.*

wind/viento aire en movimiento (90)

*El **viento** arrancó las hojas de los árboles.*

Index

Index

Index

Index

Index

Index

Index

Index

Index

Index

Index

Photo Credits